Structure and Bonding in Condensed Matter is a comprehensive introduction for those interested in understanding electronic structure as a tool for the design and synthesis of materials. The approach introduced in the book draws heavily from both the physicist's and chemist's viewpoints to establish a mathematically rigorous as well as intuitive foundation for bonding. Designed to be a self-contained treatment, the book begins with a brief introduction to necessary group theoretical concepts, followed by fundamentals of the Schrödinger equation and bonding in molecular systems. Extended systems are treated within the same framework as molecular systems, but with the benefit of group theory. All parts of the periodic table are studied with emphasis on case studies. Examples, exercises, and homework problems are used throughout the book to illustrate points and reinforce understanding.

STRUCTURE AND BONDING
IN CONDENSED MATTER

STRUCTURE AND BONDING IN CONDENSED MATTER

CAROL S. NICHOLS

Cornell University, Ithaca, NY

CAMBRIDGE
UNIVERSITY PRESS

Published by the Press Syndicate of the University of Cambridge
The Pitt Building, Trumpington Street, Cambridge CB2 1RP
40 West 20th Street, New York, NY 10011-4211, USA
10 Stamford Road, Oakleigh, Melbourne 3166, Australia

First published 1995

Printed in Great Britain at the University Press, Cambridge

A catalogue record for this book is available from the British Library

Library of Congress cataloguing in publication data
Nichols, C. S.
Structure and bonding in condensed matter / C.S. Nichols.
p.   cm.
Includes bibliographical references and index.
ISBN 0 521 46283 5 – ISBN 0 521 46822 1 (pbk)
1. Condensed matter. 2. Electronic structure. 3. Chemistry,
Physical and theoretical. 4. Chemical bonds. I. Title.
Q334.N53   1995
530.4′11–dc20   94-32183   CIP

ISBN 0 521 46283 5   hardback
ISBN 0 521 46822 1   paperback

To my Beloved

Contents

ix

Preface

Never give in, never give in, never, never, never, never – in nothing, great or small, large or petty – never give in except to convictions of honour and good sense.

Winston Churchill
Address at Harrow School, 29 October, 1941

This book grew out of a graduate course offered in the Department of Materials Science and Engineering at Cornell University. Owing to the wisdom of some of my colleagues, it was recognized that beginning graduate and advanced undergraduate students from a wide array of disciplines could benefit from a solid foundation in the electronic structure of materials. Although our students are required to take an undergraduate-level course in Solid State Physics, the materials that students learn about in that course often have little resemblance to those they encounter in their research. Indeed, it is generally the case that our students confront materials ranging from polymers to metals to ceramics and to all manner of composites therein. Furthermore, they are often involved in synthesis or characterization of these *real* materials and, as such, they contain defects. Point defects, line defects, and planar defects are topics not normally discussed in a standard Solid State Physics course.

The approach to the electronic structure of real materials used in this book borrows heavily from both the chemist's and physicist's viewpoints. Chemists have traditionally focussed more on the local chemistry of the constituent atoms, while physicists have tended to emphasize translational symmetry. Together, these two approaches provide powerful mathematical and physical rigor along with a chemist's excellent intuition about bonds. The overriding theme of the book is that one can begin from microscopic considerations such as the valence electron configuration of the constituent atoms, choose a set of suitable basis functions, and, using the insightful tools of symmetry, solve the Schrödinger equation for the energy levels and wavefunctions of this system. Such an exercise, however, is merely pedagogical if it stops there. Instead, we indicate how this process can be applied to understand electronic, optical, and mechanical properties of existing materials and ways in which new materials with altered properties might be constructed. Furthermore, the more difficult problems of systems that lack translational symmetry are tackled with the same rigor.

Structure and Bonding in Condensed Matter is designed to be used as a textbook, as each chapter is full of exercises, for example, to complete missing steps in derivations, and concludes with a set of relevant homework problems. However, it can also be used by a general reader who either has very little knowledge of or is an expert in electronic structure, since the hybrid approach is relatively new and used by only a handful of authors and researchers. The book is intended to be fairly self-contained, with additional mathematical details left to appendices. Also, it is not assumed that students are intimately familiar with the machinery of symmetry and group theory or the mathematics of quantum mechanics, and each of these topics is covered in early chapters in sufficient detail. Throughout the book, either entire chapters or sections of chapters are devoted to specific case studies that illustrate essential concepts. In these instances, material has been cited directly from the literature.

It has been argued that materials scientists and engineers have little need to know the Schrödinger equation. It is worth bearing in mind some of the success stories to come out of the approach put forward here. In one case, the students were asked to choose the best possible elemental material to be used as a getterer of phosphine gas in a silicon Molecular Beam Epitaxy system. Through careful consideration of the structure of the gas molecules and the surface of silicon, the orbitals likely to be involved in bonding, and the energetics of the bonding process, one student came up with two possible candidate materials. He chose one on the basis of availability and cost and found out from colleagues that he had chosen precisely the material they used in their system. Two other students used concepts from the course to interpret electron energy loss spectroscopy results of the local electronic structure at grain boundaries in Ni_3Al and to provide a possible explanation for the ductilizing influence of boron at such boundaries.

I encourage students and readers alike to delve into the current literature and exercise their skills on problems unfolding.

C. S. Nichols
Ithaca

Acknowledgement

This book could not have been possible without the initial and vital assistance of Ed Kramer, who made early suggestions about the content of the course; Roald Hoffmann, who was generous with his time and expertise; and Frank DiSalvo, who unselfishly shared notes, problems, and ideas. Adrian Sutton was also instrumental in early discussions about this kind of book and in encouraging me to write it. He pointed out the importance of the moments theorem and its use in nonperiodic systems. The mark of many students is on this book as well. Thank you for your patience in suffering through early drafts of these ideas. I have benefitted particularly from the comments and suggestions of Shelly Burnside, Glen Kowach, Jean Lee, David Muller, Lauri Pirttiaho, Peter Samsel, Robert Soave, Shanthi Subramanian, and Steve Townsend, among others. Elaine Park is responsible for drawing most of the figures and I thank her for her diligence. Peter Samsel is responsible for the surface plots of the hydrogen atomic orbitals; little did we realize how much work that would be. I want to thank my friend Gerald Burns, whose untimely death prevented him from seeing this in print, but I am sure he would have been pleased. Finally, David A. Smith has been a constant source of support and inspiration; I thank him for his flexibility, patience, respect, sense of humor, trust, and *je ne sais quoi*.

Physical constants and definition of some units

Physical constants

Speed of light	c	$2.997\,924 \times 10^8$ m s^{-1}
Charge of proton	e	$1.602\,189 \times 10^{-19}$ C
Charge of electron	$-e$	
Boltzmann constant	k_B	$1.380\,66 \times 10^{-23}$ J K^{-1}
Planck constant	h	$6.626\,18 \times 10^{-34}$ J s
	$\hbar = h/2\pi$	$1.054\,59 \times 10^{-34}$ J s
Electron mass	m_e	$9.109\,53 \times 10^{-31}$ kg
Proton mass	m_p	$1.672\,65 \times 10^{-27}$ kg
Bohr radius	a_0	$0.529\,177$ Å
Avogadro's number	N_A	6.022×10^{23}

Definition of some units

Å	1 Å $= 10^{-10}$ m
Electron volt	1 eV $= 1.602\,189 \times 10^{-19}$ J
Joule	1 J $= 10^7$ erg

1

Introduction

1.1 Overview

We generally tend to characterize condensed matter by its structure. Structure can mean a number of different things, of course. By *long-range structure* or *order* we mean translational symmetry: if we sit at any point inside a material, there are a huge number of other points with identical surroundings; we can get from one point in the material to any other point by simple translation. By *short-range structure* or *order* we mean that just in the vicinity of a particular atom or group of atoms does the environment look ordered. On the whole, however, the entire material does not have long-range structure. A by no means exhaustive list of materials and their attendant order includes:

- Crystalline materials are characterized by the presence of long-range order (Fig. 1.1a).

- Amorphous materials and glasses may have some short-range order, but they lack long-range order; it should be noted that some people make a distinction between glasses and amorphous materials while others do not, based on the existence of a glass transition temperature (Fig. 1.1b).

- Quasicrystals are an interesting mixture of the above two cases; they have short-range order, but have long-range structure that is *quasiperiodic*, but not periodic (Fig. 1.1c).

- Liquids may have short-range order, such as is found in the case of water, where hydrogen bonds play an important role, but they lack long-range order (Fig. 1.1d).

- Polycrystalline materials have regions of crystalline order inside grains, but also have regions which may be disordered at grain boundaries; overall they lack long-range order because of differences in orientation of grains (Fig. 1.1e).

1

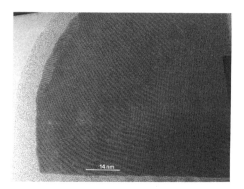 

Fig. 1.1(a). High-resolution transmission electron micrograph of bulk platinum (left). Note the presence of lattice fringes. On the right is a diffraction pattern of the same sample. The sharp spots indicate that the sample is crystalline in nature. Figures courtesy of Yung-Cheng Lu.

- Polymers have short-range order if we look at the orientation of atoms along the polymer backbone; the local environment around each carbon atom is approximately identical. Crystalline polymers also have translational order and there are other phases that are somewhere in between (Fig. 1.1f).
- Alloys may be structurally ordered, but chemically disordered. Think of a simple cubic material populated with A and B atoms: if both the structure and the A and B atoms are ordered, then the material has long-range order; if the structure is ordered, but the A and B atoms are randomly arranged on the lattice sites, then the material is still structurally ordered, but chemically disordered (Fig. 1.1g).
- Artificial structures comprise a large variety of structures from superlattices to composites and they may be ordered in both the long and short range. For example, semiconductor superlattices may be a consistently repeating unit cell, but composites other than ideal laminates lack long-range order (Fig. 1.1h).

A physicist or materials scientist might look at a material, describe its long-range structure, and declare that it will have certain properties. This is an example of one of the guiding principles behind materials science. That is, if we can discern structure–property relationships for simple systems, then we can use this information to design, at least in principle, new materials. Bonding, in the sense of where the electrons are and what the electronic structure of the material is, is considered almost indirectly or as an afterthought. A chemist, on the other hand, will look at the constituents

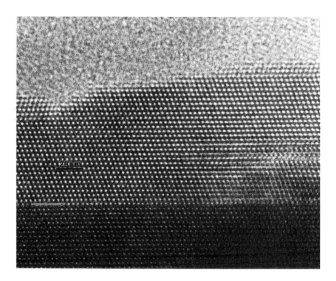

Fig. 1.1(b). High-resolution electron micrograph of a NiSi$_2$ precipitate (bottom) in contact with a crystalline silicon region (middle) which is in contact with an amorphous silicon region (top). Note the difference between the amorphous and crystalline regions; the former lacks the long-range order of the latter. From C. Hayzelden and J.L. Batstone, *J. Appl. Phys.* **73**, 8279 (1993). Used by permission.

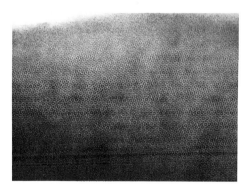 

Fig. 1.1(c). High-resolution electron micrograph (left) of the quasicrystalline material Al$_{63}$Cu$_{23.5}$Fe$_{13.5}$ (atomic percent). The accompanying diffraction pattern (right) shows clearly the quasicrystalline symmetry of this sample, particularly the ten-fold symmetry of spots. Figure courtesy of C. Hayzelden.

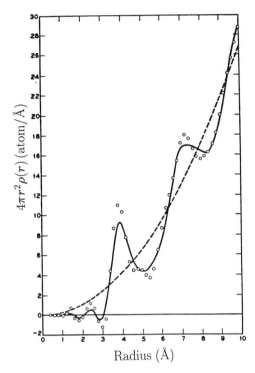

Fig. 1.1(d). The radial distribution function for liquid argon. This function gives the probability of finding an atom at a radius *r* from an arbitrary atom in the liquid. Sharp peaks, such as are found in a crystalline sample, indicate well-defined and regular atom spacings. The absence of such peaks, as seen in this figure, indicate a loss of translational symmetry. The solid line was measured by D.G. Henshaw, *Phys. Rev.* **105**, 976 (1957); open circles are from B.A. Dasannacharya and K.R. Rao, *Phys. Rev.* **137**, A417 (1965); and the dashed line is the average atomic density. Used with permission from H.L. Frisch and Z.W. Salsburg, *Simple Dense Fluids* (Academic Press, New York, 1968), p. 151.

of a material and from their chemistry (by which we mean here the valence electron configuration) declare that the material will have a certain type of bonding and that this bonding dictates the structure. The properties of the material are evaluated from the local, atomic perspective. A schematic of these two points of view is presented in Fig. 1.2.

   Do we view condensed matter from the atoms up or from the macroscopic sample down? Clearly these approaches are not independent. One's point of view merely dictates which length scale gets precedence over the other. We will attempt to borrow from both points of view and, in some sense, to synthesize the two to understand the relationship among the constituents of a

Fig. 1.1(e). Transmission electron micrograph of polycrystalline gold. The differences in shades from one grain to the next are caused by changes in relative orientation with respect to the electron beam. The scale marker represents 100 nm. Figure courtesy of D.A. Smith.

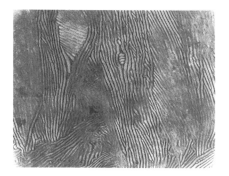

Fig. 1.1(f). Electron micrographs of lamellae of high density polyethylene. A small selected area of this micrograph is deceptive as it looks ordered locally, but clearly lacks long-range order. Magnification is ×78 000. From D.T. Grubb and A. Keller, *J. Polymer Science* **18**, 207 (1980). Used with permission.

material and their chemistry, their bonding and electronic structure, and the structure of the material. Our primary goal is to understand the electronic structure of existing materials and to be able to design new materials with desirable properties from electronic structure considerations.

In order to carry out this task, we must first gather a number of tools that will provide us with the basis for discussing the structure of solids systematically as well as for developing a theory of bonding. The first tool we require is symmetry and some rigorous way of using it. We will discover

Fig. 1.1(g). Electron diffraction pattern from polycrystalline chemically disordered Cu_3Au (left). The dots in each ring correspond to an allowed fcc reflection. Electron diffraction pattern from a single grain of chemically ordered Cu_3Au which has a simple cubic lattice. The extra fainter dots arise from the chemical ordering (superlattice reflections). Figures courtesy of D.A. Smith.

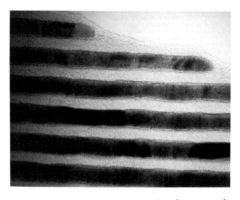

Fig. 1.1(h). High-resolution electron micrograph of a superlattice of Au/Al_2O_3. The bilayer repeat distance is 110 Å. Figure courtesy of C. Hayzelden.

that long-range order forms the cornerstone of electronic structure theory. But for many classes of materials, there is no translational symmetry and we need to exploit short-range order. The theory of finite groups provides a framework for discussing both short-range and long-range order. The first few chapters develop the concepts of group theory in the context of using it for understanding electronic structure.

We also need to understand the building blocks of the solid state. These are atoms, and the behavior of atoms is explained using quantum mechanics. We therefore spend some time reviewing essential mathematics, discussing the partial differential equation that codifies quantum mechanics –

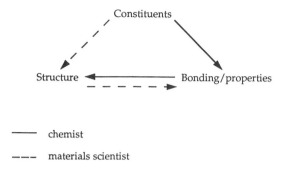

——— chemist

——— materials scientist

Fig. 1.2. Schematic diagram showing the interplay between structure, constituents, and bonding/properties. The arrows suggest the direction of investigation favored by chemists (solid lines) and by materials scientists (dashed lines).

the Schrödinger equation – and the solution of this equation for two simple atomic systems. Both the hydrogen atom and a generic diatomic molecule offer a great deal of insight into electronic structure and bonding that can be carried over to more complicated solids.

Once we have mastered symmetry and elementary quantum mechanics, it is time to combine the two and explore the dominant role that long-range order plays in determining the electronic structure of materials. A number of useful concepts will be introduced and developed within the context of simple one-dimensional systems. We will also investigate tools necessary for when our system lacks translational symmetry.

After generalizing these tools from one dimension to two and three dimensions, we will spend a brief chapter discussing crystal structure.

Finally, then, we are in a position to examine materials from all parts of the periodic table. We will cover explicitly some case studies in which the valence electrons are localized in the region around and between atoms and materials at the opposite extreme – free-electron and nearly-free-electron systems. We round out the discussion by considering the transition metals, for which the *d* electrons are important.

We conclude by examining briefly more sophisticated methods of calculating the electronic structure of real materials than we have been able to cover in this Introduction. Also, we examine some of the assumptions that we have made about our systems and how these assumptions might be relaxed. At the close of the book, the reader will be knowledgeable enough to delve into the current literature.

1.2 The periodic table

We know that trends exist in the properties of elements: *e.g.*, chemical valency, electrical conductivity, *etc.* We would like, in our quest for understanding materials and in our search for new materials, to be able to have a systematic foundation of knowledge. Rutherford once complained that 'chemistry was stamp collecting', meaning that there appeared to be little that was systematic in chemistry. We need to be able to refute this claim and we will begin by describing attempts to classify the (known) elements.

Division of matter into various forms is an ancient concept, dating back to Aristotle. The development of the modern form of the periodic table, however, begins in the early 19th century. It is a rich story in which many prominent scholars, including Goethe, figure. The modern periodic table of the elements is a reflection of atomic theory. Early classification attempts, ignorant of such theory, were largely empirical and were based on atomic weight and chemical reactivity.

A German chemist, and colleague of Goethe at Jena, by the name of Johann Wolfgang Döbereiner was the first to report a relationship between atomic weights of elements with analogous properties. In 1817, he noted that there were triads of elements, such as (Cl,Br,I) and (Co,Sr,Ba). When arranged in order of increasing atomic weight, the middle elements appeared to have properties that were roughly the average of the outer two. In the years following Döbereiner's announcement, others identified new triads and, in some cases, groups of four or five elements.

In 1863, John Newlands, a chemist then working in London, carried Döbereiner's idea further by arranging all the 56 then known elements in order of increasing weight. He pointed out that the properties of the eighth element resembled those of the first, those of the ninth element resembled those of the second, and so on. He cited a 'law of octaves' by analogy with music for his choice of groups of eight. While some columns contained elements with like properties, there were some drawbacks to Newlands' scheme. Namely, there was no room left for new elements and some elements did not seem to quite fit in. For example, Be, Al, and Zr were classified together. At a meeting of The Chemical Society in London (1865), Newlands was subjected to a great deal of ridicule by his colleagues. One went so far as to suggest that a better classification scheme might be based on the first letter of the elements, for surely any scheme would yield occasional coincidences.

The correct modern periodic table arrived shortly thereafter, however. In 1869, the Russian chemist Dmitri Mendeleev and, simultaneously, the German Lothar Meyer followed Newlands' idea of arrangement by increasing

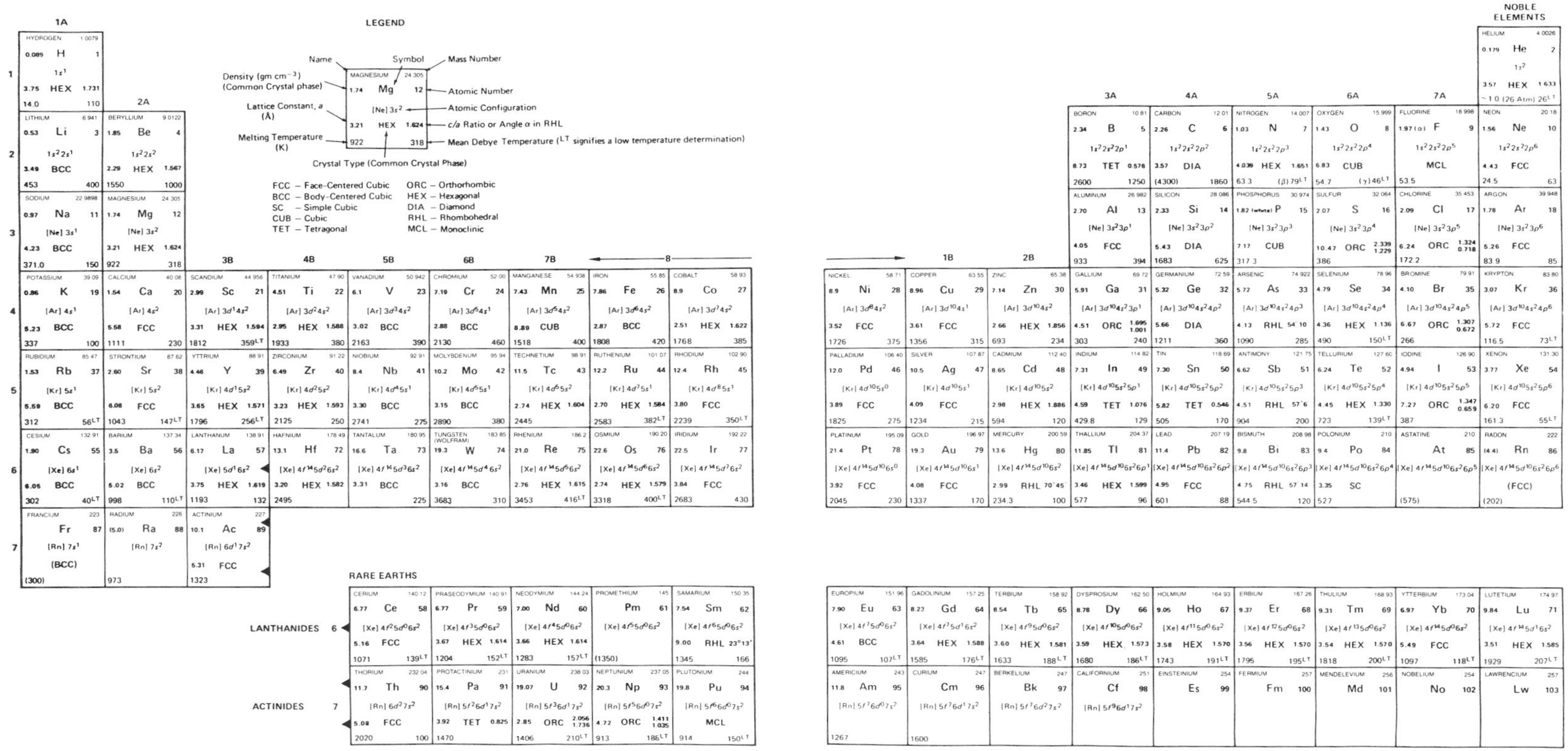

Fig. 1.3. The modern periodic table of the elements. Reprinted with permission from N.W. Ashcroft and N.D. Mermin, *Solid State Physics* (Saunders College, Philadelphia, 1976).

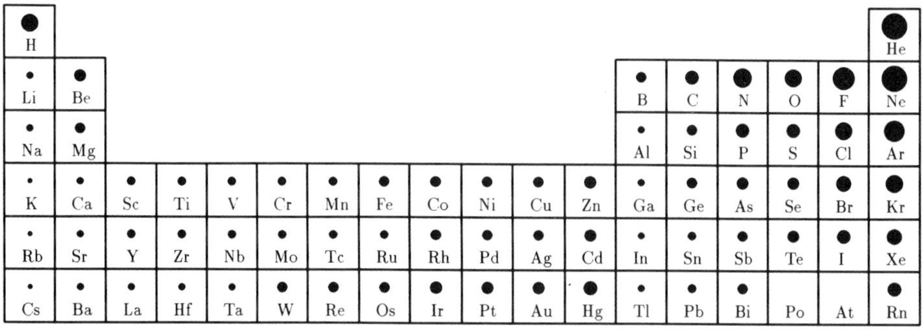

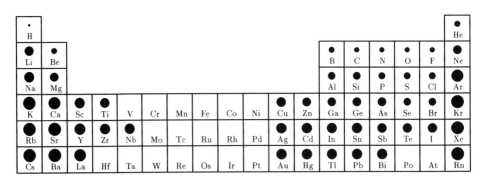

Fig. 1.4. First ionization energies of the elements, excluding the lanthanides and actinides. The larger the circle, the larger the ionization energy. The maximum is He. After R.T. Sanderson, *Chemical Periodicity* (Reinhold Publishing Corporation, New York, 1960), p. 22.

Fig. 1.5. Nonpolar covalent radii of the elements, excluding the lanthanides and actinides. The larger the circle, the larger the radius. After R.T. Sanderson, *Chemical Periodicity* (Reinhold Publishing Corporation, New York, 1960), p. 26.

atomic weight. However, a number of improvements were implemented. Firstly, long periods were instituted for elements known as transition metals. Secondly, blank spaces were left if known elements did not fit in. Following from the blank spaces, a few predictions of new elements were made. One example is that of 'ekasilicon'; the element directly below Si was not known at the time of the original table, but it was subsequently discovered and named 'germanium'. Since then, of course, many other elements have been discovered and inserted into the table.

The modern periodic table, Fig. 1.3, is grossly characterized into three main groups:

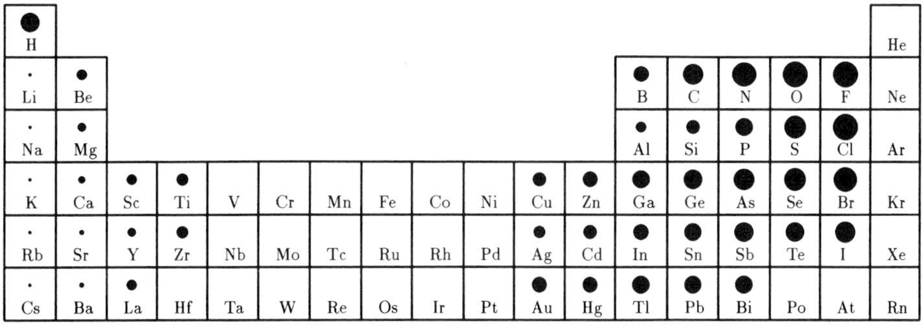

H																	He
Li	Be											B	C	N	O	F	Ne
Na	Mg											Al	Si	P	S	Cl	Ar
K	Ca	Sc	Ti	V	Cr	Mn	Fe	Co	Ni	Cu	Zn	Ga	Ge	As	Se	Br	Kr
Rb	Sr	Y	Zr	Nb	Mo	Tc	Ru	Rh	Pd	Ag	Cd	In	Sn	Sb	Te	I	Xe
Cs	Ba	La	Hf	Ta	W	Re	Os	Ir	Pt	Au	Hg	Tl	Pb	Bi	Po	At	Rn

Fig. 1.6. Electronegativities of the elements, excluding the lanthanides and actinides. The larger the circle, the larger the electronegativity. After R.T. Sanderson, *Chemical Periodicity* (Reinhold Publishing Corporation, New York, 1960), p. 32.

(i) representative elements: metals/nonmetals with a great variety of properties;

(ii) transition metals: metals that are similar to one another, yet different from the representative metals;

(iii) inner transition metals: lanthanides and actinides.

The periodic table is particularly useful for identifying trends in various properties. It is important to become familiar with these properties and how they vary throughout the periodic table since we will make extensive use of them. The first ionization energy is the amount of energy required to remove the outermost electron from an element. This quantity generally increases across a given row and decreases down a given column. See Fig. 1.4. The atomic size of an element is not an exactly defined quantity, but it can be thought of as the radius at which the electron density drops to a small value. There is no smooth trend all the way across a row or column, although there is a general increase down a column. See Fig. 1.5. Electronegativity is a valuable concept when we want to consider the ionicity of a bond, since, according to Pauling, it is the 'relative' attraction of an atom for the valence electrons in a covalent bond. If two atoms attract the electrons in a bond equally, their electronegativities are equal and the bond is *nonpolar*. For two atoms that do *not* share the electrons in a bond equally, the bond is *polar*. The element that attracts electrons more readily has the higher electronegativity. Electronegativity peaks at F, in the upper right of the periodic table, because it has an outer shell of 7 electrons. Addition of one more electron completes the shell and, hence, is very stable. Electronegativity decreases down a column and across a row to the left. See Fig. 1.6.

2

Group theory: background and basics

2.1 Introduction

All objects, we will discover, have some symmetry. That is, there is some *symmetry operation* that we can apply to the object so that it is identical in appearance both before and after application of the symmetry operation. Why, you may ask, do we care about symmetry operations when our goal is to learn about the electronic structure of materials? There are a number of motivations. Firstly, some of you may already know that the electronic structure, or band structure, of a solid is expressed in the space of the first Brillouin zone, which contains all the symmetry of the crystal. Some examples of common first Brillouin zones, for a bcc lattice and an fcc lattice, are shown in Fig. 2.1. The calculated band structure of Na, which has a bcc structure, is shown in Fig. 2.2. Notice how the symbols that appear in the Brillouin zone for a bcc lattice in Fig. 2.1 appear in Fig. 2.2. One nice thing that symmetry does for us is that it reduces the amount of the Brillouin zone over which we need to calculate the band structure: to do all of the zone is redundant, since various parts are related by symmetry. The band structure of Si in the diamond structure (fcc lattice) is shown in Fig. 2.3. Notice how the symbols that appear in the Brillouin zone for an fcc lattice in Fig. 2.1 appear in Fig. 2.3. There are a relatively small number of Brillouin zones and therefore we will only need to become familiar with the symmetry of a few objects.

Some examples of symmetry operations are rotations, reflections, inversions or simply doing nothing to the object. As a specific example, consider the benzene molecule, C_6H_6, shown in Fig. 2.4. The molecule is planar, and we have neglected to show explicitly the H atoms attached to each C atom. One striking thing about the benzene molecule is the presence of an axis of rotation about which we can rotate the molecule up to six times before

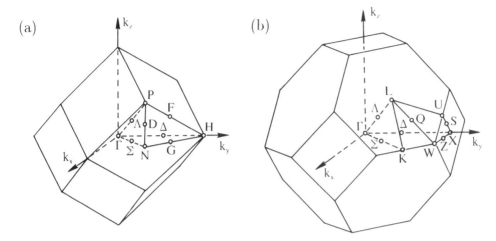

Fig. 2.1. Brillouin zone for (a) a bcc lattice and (b) an fcc lattice. See the text for further explanation.

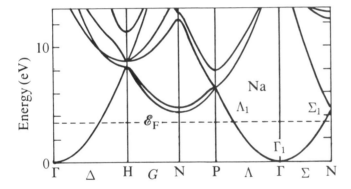

Fig. 2.2. The calculated band structure of metallic crystalline Na around a circuit of special points in the Brillouin zone. Some of the bands are labeled.

it returns to its initial configuration. This axis is denoted z in the figure. Also note that there are a number of rotation axes that are perpendicular to z. These are given the various symbols C_2^d and C_2'. There are a number of mirror planes as well. A mirror plane in general is given the symbol σ. To indicate that it has some special property a subscript is attached, such as σ_d or σ_v. We should not forget that the benzene molecule also has the symmetry operation in which we do nothing to the molecule. This is known as the identity and is given the symbol E. Finally, the molecule has more

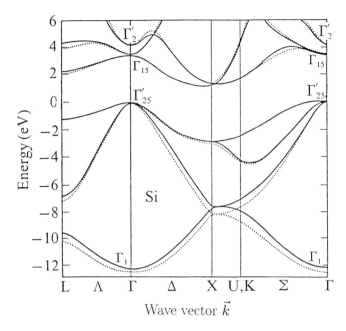

Fig. 2.3. The calculated band structure of crystalline silicon in the diamond structure around a circuit of special points in the Brillouin zone. Some of the bands are labeled.

complicated operations called *improper rotations*. An improper rotation is a rotation followed by a reflection; it is denoted S_n.

We will show later in this chapter that the collection of symmetry operations of any object forms a group in the mathematical sense. First, however, we familiarize ourselves with individual symmetry operations and some of their properties.

2.2 Point symmetry operations

What we have not stated explicitly in examining the symmetry operations of the benzene molecule is that one point of the molecule has always remained fixed. This gives us the so-called *point symmetry operations*. Later we will add translational symmetry and find even more symmetry operations.

In noncubic crystals and for molecules, we can usually find one axis that has higher rotational symmetry than the others. This axis is defined as the *principal axis* and it is always labeled as the *z* axis. For orthorhombic crystals, there is a certain arbitrariness about what we call the principal axis.

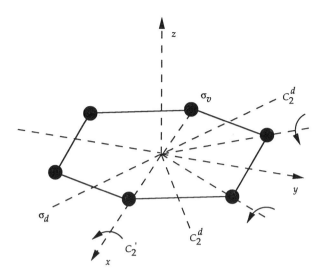

Fig. 2.4. Diagram of a planar benzene (C_6H_6) molecule with some of the symmetry operations indicated.

For cubic crystals, we can see that there are three four-fold rotation axes; hence, we can choose any one of these as the principal axis.

There are two different widely used notations for symmetry operations. One is called the Schoenflies notation and the other is called the International Notation. We will always use the Schoenflies notation, for reasons of pure convenience. One notation does not contain any more information than the other.

An *exhaustive* list of all point symmetry operations is given in Table 2.1. *We will adhere to the convention of performing rotations in a counterclockwise fashion when looking* down *the z axis (i.e., from positive z toward negative z).* Note that there is a difference between a *symmetry operation* and a *symmetry element*. A *symmetry operation* is one of the items listed above and these all involve movement of the body. A *symmetry element* is a geometrical entity possessed by the body, such as an axis, a plane, or a center of inversion. Some specific examples of the complete set of symmetry operations of a few selected molecules are shown in Fig. 2.5. A final word about notation: proper or improper rotations about the principal axis are given no other notation except subscripts when appropriate; rotations perpendicular to the principal axis are primed or given another distinguishing superscript; reflections are also sometimes given superscripts to distinguish between two different planes. *There are no hard and fast rules, however!* Also, beware that some people

Table 2.1. *An exhaustive list of point symmetry operations*

Operation symbol	Operation	Comments
E	Identity	The object is not rotated at all or rotated by 2π. All objects possess this operation.
i	Inversion	The object is inverted through an origin which is the center of inversion.
C_n	Rotation	This operation is also known as a proper rotation. It is a rotation of the object counterclockwise by $2\pi/n$ about the principal axis.
σ	Reflection	Reflection of the object across a mirror plane; all points from one side of the plane are moved to an equidistant position on the other side of the plane and *vice versa*.
σ_h	Horizontal reflection	Reflection through a plane that is perpendicular to the principal axis.
σ_v	Vertical reflection	Reflection through a plane that contains the principal axis.
σ_d	Diagonal reflection	Reflection through a plane that contains the principal axis and bisects the angle between two-fold axes perpendicular to the principal axis.
S_n	Improper rotation	A rotation by $2\pi/n$ counterclockwise about the principal axis followed by a horizontal reflection; $S_n = \sigma_h C_n$.

may choose origins differently; this can lead to a difference in notation as well. We will generally try to always select the origin as the point of highest symmetry.

Exercise 2.1   For the Platonic solids shown in Fig. 2.6, work out the complete set of symmetry operations.

In all of the examples discussed above, for each symmetry operation there is a corresponding operation that takes the object back to its original position. Mathematically,

$$AB(\text{object}) \; = \; E(\text{object}) \tag{2.1}$$

where A and B are symmetry operations and clearly B is the inverse of A. Usually the order of application of symmetry operations is important. The

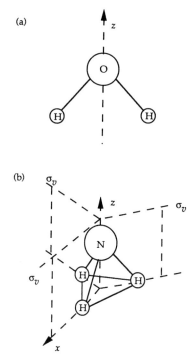

Fig. 2.5. Diagrams of various molecules that contain some of the point symmetry operations described in the text. (a) H_2O and (b) NH_3.

operation that appears furthest to the right is applied to the object first, followed by any subsequent operations. But we will see that order is of no consequence for inverses. As an example, for the H_2O molecule, the inverse operation for C_2 is C_2 itself. Verify this for yourself.

Occasionally (quite often, actually), we will want to take the product of two symmetry operations. As noted above for inverses, the operations are applied successively, and can be denoted by, *e.g.*, $C_3C_3 = C_3^2$. Clearly the product of three C_3 rotations, $C_3C_3C_3$, can be written symbolically as C_3^3. This product is equal to the identity, E.

2.3 Stereographic projection

We can see that to represent symmetry operations in any sort of compact way can be a cumbersome task. However, a useful and concise way of visualizing the effects of symmetry operations on an object is by stereographic projection.

Tetrahedron

Octahedron

Icosahedron

Cube

Dodecahedron

Fig. 2.6.   The five Platonic solids. Used with permission from D. Schattschneider and W. Walker, *M.C. Escher Kaleidocycles* (Pomegranate Artbooks, Inc., Corte Madera, CA, 1977), p. 6.

Definition 2.1 *A point in the +z hemisphere is projected onto the xy plane by finding the intersection between this plane and a line drawn between the chosen point and the south pole of the unit sphere. A dot is placed at the location of this intersection.*

As an example, a point along the $+z$ axis is projected onto the center of a unit circle in the xy plane. A point in the $-z$ hemisphere is projected using the north pole and an '$\times$' or an open dot ,'$\circ$', is used to mark the intersection with the xy plane.

The procedure used to generate a stereographic projection of an object is as follows:

(i) Choose an arbitrary point; this is called a *general equivalent position* (gep). We choose a gep because only such points will display the full collection of symmetry operations.
(ii) Apply each independent symmetry operation to the gep and map the resultant position with a dot or open dot, as appropriate.
(iii) There will always be as many points as independent symmetry operations.
(iv) The final stereographic projection displays the projections of the gep under all symmetry operations in the group. We can also include in this final diagram the symmetry elements of the group. The symbols for the various symmetry elements are shown in Table 2.2.

As an example, consider the collection of symmetry operations of the H_2O molecule: $\{E, C_2, \sigma_v, \sigma_v'\}$. E maps the arbitrary point '1' onto itself. See Fig. 2.7. C_2 maps '1' onto point '2'. The first mirror reflection, σ_v, takes '1' onto point '3', while the second mirror reflection, σ_v', takes '1' onto point '4'. Note that the number of points generated equals the number of independent symmetry operations, in this case, four.

Exercise 2.2   Using the collection of symmetry operations that you found for the tetrahedron (Exercise 2.1), construct the stereographic projection.

2.4 32 crystallographic point groups

Molecules have no constraints on 'n' for C_n rotations or S_n improper rotations, so that there is no limit on the number of allowed point groups. Remember, a group for us right now is just a collection of symmetry operations. However, crystals have to pack space, and the constraint of translational symmetry limits n to 1, 2, 3, 4, and 6. This means that there are only

Table 2.2. *Symbols for the various symmetry elements and their meanings*

Symbol	Meaning
———	Line of reflection.
⌐ˎ ˋ .	Line of glide–reflection. The half–arrowheads indicate the size of the translation (glide) associated with the reflection in the dashed line.
⟨	Center of two-fold rotation (reflection in a point).
△	Center of three-fold rotation.
□	Center of four-fold rotation.
◇	Center of six-fold rotation.
❘	Center of two-fold rotation lying on a line of reflection.
▲	The corresponding center for three-fold rotation.
■	The corresponding center for four-fold rotation.
⬣	The corresponding center for six-fold rotation.

32 distinct groups of symmetry operations; these are called *crystallographic point groups.*

You have probably been told *ad nauseam* that objects with five-fold symmetry cannot tile the plane (by 'tile' we mean that a consistent set of objects cannot be used to cover completely the plane without overlapping the objects or leaving gaps between them). A clever little proof of this follows.

Consider two lattice points A and A′ separated by a unit translation vector, $\vec{t}$, as shown in Fig. 2.8. For 'lattice point', imagine a point that sits at the intersection of the lines on a piece of square graph paper. For 'translation vector', take the shortest vector on the same piece of graph paper; that is, the vector from one lattice point to a neighboring point. Apply a rotation by an angle α to AA′ and an equal rotation in the opposite direction. Now, if B and B′ are to be lattice points, then $|\vec{t'}|$ must be an integral number times $|\vec{t}|$:

$$| \vec{t'} | = m | \vec{t} |\tag{2.2}$$

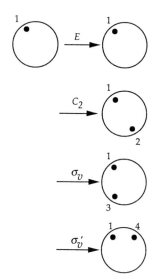

complete collection:

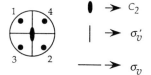

Fig. 2.7. Stereographic projection of the symmetry operations for the group of the H_2O molecule.

From the diagram

$$| \vec{t}' | = -2 | \vec{t} | \cos \alpha + | \vec{t} | \qquad (2.3)$$

and substituting Eq. 2.3 into Eq. 2.2,

$$\cos \alpha = \frac{(1 - m)}{2} = \frac{M}{2} \qquad (2.4)$$

where clearly if $(1 - m)$ is an integer, then M must also be an integer. Now

$$| \cos \alpha | \leq 1 \qquad (2.5)$$

or

$$| M | \leq 2 \qquad (2.6)$$

which has solutions $M = -2, -1, 0, 1, 2$. This means that α can take on values $\pi, 2\pi/3, \pi/2, \pi/3, 0$. Or, in terms of $2\pi/n$, $n = 2, 3, 4, 6$, and 1.

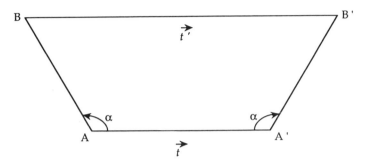

Fig. 2.8.   Geometrical diagram accompanying proof that objects of five-fold symmetry cannot tile the plane.

As an aside, it is interesting to inquire how to pack *space*, rather than just tiling the plane. The search for solids that pack space is a very old problem, dating back to the mathematician Plateau (1873) and the physicist Kelvin (1887). In turns out that there are five objects that will fill all of space without leaving any empty space between the objects or without causing overlaps. These are: the triangular, rhombic and hexagonal prisms, the rhombic dodecahedron, and the tetrakaidecahedron (the last object is sometimes referred to as 'Kelvin's tetrakaidecahedron'). The packing of these polyhedra to fill space is shown in Fig. 2.9.

How do we set about finding the 32 crystallographic point groups? We build up the catalog of point groups by successively adding on additional symmetry elements. For example, take an object that has no symmetry elements. This simplest group is $\{E\}$. Next add on a rotation axis of order 2. The group thus generated is $\{E, C_2\}$, and so on for higher rotation axes. These groups possess the point symmetry $C_1, C_2, C_3, \ldots$ or we say that they transform as the C_n point group. As a warning: do not confuse the symmetry operation C_n with the point group notation. It will generally be clear from the context.

The next level of complexity comes when we add a horizontal mirror plane. The new point groups are given the notation C_{nh}.

Exercise 2.3   Take the group C_3, which has symmetry operations $\{E, C_3, C_3^2\}$. Add σ_h to obtain the group C_{3h}, which has six elements. Determine the six elements.

We can continue the process of generating the 32 point groups by adding other symmetry elements, σ_v, *etc.* All of the crystallographic point groups are

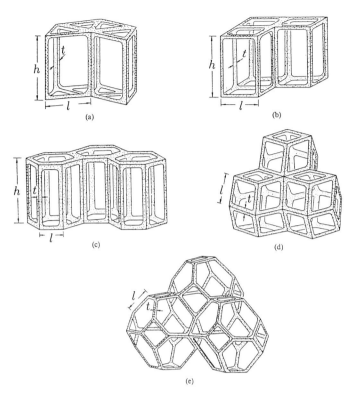

Fig. 2.9. Polyhedra that pack to fill three-dimensional space: (a) triangular prisms, (b) rectangular prisms, (c) hexagonal prisms, (d) rhombic dodecahedra, and (e) tetrakaidecahedra. From L.J. Gibson and M.F. Ashby, *Cellular Solids, Structure and Properties* (Pergamon Press, Oxford, 1988), p. 23. Used with permission.

listed in Table 2.3. The stereographic projections for all of the point groups are listed in Appendix 2.

For the 32 crystallographic point groups, there can be at most 48 symmetry operations. The group that has all these operations is the octahedral group, O_h.

Given that we now know what the 32 crystallographic point groups are, how do we classify an object into one of these groups? The following is a guideline to be used for determining the point group of an object. Since the cubic point groups are characterized as having no unique principal axis, and four C_3 axes, they are more difficult to categorize systematically.

For noncubic objects, follow the scheme outlined in Fig. 2.10. Two examples will suffice to demonstrate its use. Consider first the H_2O molecule, which has symmetry operations $\{E, C_2, \sigma_v, \sigma'_v\}$. From Fig. 2.10, the point

Table 2.3. *The 32 crystallographic point groups*

Schoenflies notation	International notation	Symmetry operations
Triclinic		
C_1	1	E
S_2	$\bar{1}$	E, i
Monoclinic		
C_2	2	E, C_2
C_{1h}	m	E, σ_h
C_{2h}	$2/\mathrm{m}$	E, C_2, i, σ_h
Orthorhombic		
D_2	222	E, C_2, C_2', C_2''
C_{2v}	mm2	$E, C_2, \sigma_v, \sigma_v'$
D_{2h}	mmm	$E, C_2, C_2', C_2'', i, \sigma_h, \sigma_v, \sigma_v'$
Tetragonal		
C_4	4	$E, 2C_4, C_2$
S_4	$\bar{4}$	$E, 2S_4, C_2$
C_{4h}	$4/\mathrm{m}$	$E, 2C_4, C_2, i, 2S_4, \sigma_h$
D_4	422	$E, 2C_4, C_2, 2C_2', 2C_2''$
C_{4v}	4mm	$E, 2C_4, C_2, 2\sigma_v, 2\sigma_d$
D_{2d}	$\bar{4}$2m	$E, C_2, 2C_2', 2\sigma_d, 2S_4$
D_{eh}	$4/\mathrm{mmm}$	$E, 2C_4, C_2, 2C_2', 2C_2'', i, 2S_4, \sigma_h,$ $2\sigma_v, 2\sigma_d$
Trigonal		
C_3	3	$E, 2C_3$
S_6	$\bar{3}$	$E, 2C_3, i, 2S_3$
D_3	32	$E, 2C_3, 3C_2$
C_{3v}	3m	$E, 2C_3, 3\sigma_v$
D_{3d}	$\bar{3}$m	$E, 2C_3, 3C_2, i, 2S_6, 3\sigma_v$
Hexagonal		
C_6	6	$E, 2C_6, 2C_3, C_2$
C_{3h}	$\bar{6}$	$E, 2C_3, \sigma_h, 2S_3$
C_{6h}	$6/\mathrm{m}$	$E, 2C_6, 2C_3, C_2, i, 2S_3, 2S_6, \sigma_h$
D_6	622	$E, 2C_6, 2C_3, C_2, 3C_2', 3C_2''$
C_{6v}	6mm	$E, 2C_6, 2C_3, C_2, 3\sigma_v, 3\sigma_d$
D_{3h}	$\bar{6}$m2	$E, 2C_3, 3C_2, \sigma_h, 2S_3, 3\sigma_v$
D_{6h}	$6/\mathrm{mmm}$	$E, 2C_6, 2C_3, C_2, 3C_2', 3C_2'', i,$ $2S_3, 2S_6, \sigma_h, 3\sigma_v, 3\sigma_d$
Cubic		
T	23	$E, 8C_3, 3C_2$
T_h	m3	$E, 8C_3, 3C_2, i, 8S_6, 3\sigma_h$
O	432	$E, 8C_3, 3C_2, 6C_2', 6C_4$
T_d	$\bar{4}$3m	$E, 8C_3, 3C_2, 6\sigma_d, 6S_4$
O_h	m3m	$E, 8C_3, 3C_2, 6C_2, 6C_4, i, 8S_6,$ $3\sigma_h, 6\sigma_d, 6S_4$

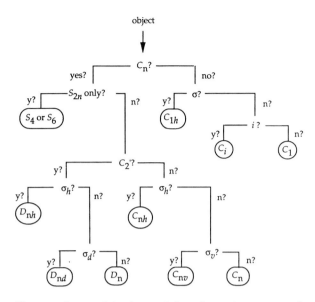

Fig. 2.10. Chart to be used in determining the point group of an object.

group is C_{2v}. Next consider the ammonia molecule, NH_3. It has symmetry operations $\{E, C_3, C_3^2, \sigma_v^1, \sigma_v^2, \sigma_v^3\}$, so that the point group is C_{3v}. The superscript on the σ operations represents different mirror planes.

If we now look at the character tables in Appendix 3 (which we will get to understanding in Ch. 3), then we can locate the point groups C_{2v} and C_{3v} and find the symmetry operations they contain. We will find that they are identical with those found for H_2O and NH_3, respectively. Note for the point group C_{3v} that C_3 and C_3^2 are lumped together. These symmetry operations are said to be in the same 'class', which we will come to in section 2.8.3.

2.5 Other symmetry operations

We have only considered symmetry operations about a fixed point. However, we know that crystals have translational symmetry which can be expressed as:

$$\vec{t}_n = n_1 \vec{a}_1 + n_2 \vec{a}_2 + n_3 \vec{a}_3 \qquad (2.7)$$

where n_1, n_2, and n_3 are integers and $\vec{a}_1, \vec{a}_2$, and $\vec{a}_3$ are noncoplanar vectors describing the edges of the unit cell that packs space; $\vec{t}_n$ is a symmetry operation.

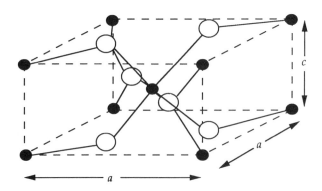

Fig. 2.11. A unit cell of MO_2 in the rutile structure, where M is a transition metal atom. The oxygen atoms are the large open circles and the transition metal atoms are the small solid circles.

Suppose that we combine rotation or reflection symmetry operations with translational symmetry. We will generate the kinds of operations that can be found in a crystal. These are known as glides and screws. A *screw operation* is a rotation followed by a translation parallel to the axis of rotation. A *glide operation* is a reflection followed by a translation.

An example that will arise again when we discuss its electronic structure is MO_2 in the rutile structure, where M is one of the transition metals Ti, Re, Rh, *etc.* This is shown in Fig. 2.11. If we treat the unit cell just as a molecule, then we find that the point group is D_{2h}. However, consider a C_4 rotation about the c axis followed by a translation of $c/2$ along $\hat{c}$ and $a/2$ along the perpendicular axes. This operation, which is a screw symmetry operation, leaves the crystal in an equivalent position and, hence, it must be a symmetry operation of the crystal.

The combination of point operations with translations is given the notation

$$\{C_4 \mid \vec{\tau}\} \tag{2.8}$$

where $\vec{\tau} = \frac{1}{2}\vec{a} + \frac{1}{2}\vec{a} + \frac{1}{2}\vec{c}$. The first symbol refers to the type of point operation, while the second refers to a translation. Note that the translation must also be specified for a complete description of the operation.

Just as a point group is the collection of all point symmetry operations of an object, a *space group* is the set of symmetry operations $\{R \mid \vec{\tau}\}$ which transforms the crystal into itself. A general point symmetry operation is given the symbol 'R'. It turns out that there are 230 independent space groups, identified in much the same way as are the point groups.

If a space group contains no glide or screw operations, then it is a *symmorphic* space group. The point group of these space groups can be generated according to our scheme above.

If a space group contains glide or screw operations, then it is a *nonsymmorphic space group*. There exists a procedure for obtaining the point group of the space group in this case as well, but we will not discuss it here.

2.6 Group definition

Up to this point we have considered the symmetry operations of an object as a simple collection. We will now find out that they actually constitute a group in the formal mathematical sense. There is a very rich structure to this relationship.

A mathematical *group* is a collection of elements that obey the following four postulates, in addition to possessing a multiplication operation for interrelating the elements. The operation may be addition, multiplication in the ordinary sense, matrix multiplication, or multiplication of symmetry operations such as we have discussed above.

(i) Closure: the result of multiplying any two elements in the set must also be in the set.

(ii) Identity: one of the elements of the set must be the identity, such that $EA_j = A_j E = A_j$ for all A_j in the set.

(iii) Inverse: every element in the set must possess an inverse, which is also in the set. $A_i^{-1} A_j = E$, where $A_j^{-1} = A_i$ and A_j is a member of the set.

(iv) Associativity: $(AB)C = A(BC)$. Do not confuse this with commutativity, which concerns the order of applying the elements.

If the elements of the set obey (i)–(iv), then the set forms a group. A group may have a *finite* or an *infinite* number of elements. If the group of elements is finite, then the number of elements is called the *order* of the group. We will not deal with infinite groups.

Consider the example of the collection of elements $\{1, -1\}$. Is this set a group? The operation is ordinary multiplication. Closure is satisfied; we cannot generate an element outside of the set. The identity element is 1. Each element is its own inverse and associativity is clearly obeyed. This set does form a group.

Exercise 2.4   {all integers, including zero}. The operation is addition. Verify that this collection forms a group.

Exercise 2.5 $\{E, C_2, \sigma_v, \sigma_v'\}$ is the collection of symmetry operations of the water molecule. Verify that this is a group.

Table 2.4. *Multiplication table for the point group* C_{2v}

C_{2v}	E	C_2	σ_v	σ_v'
E	E	C_2	σ_v	σ_v'
C_2	C_2	E	σ_v'	σ_v
σ_v	σ_v	σ_v'	E	C_2
σ_v'	σ_v'	σ_v	C_2	E

2.7 Multiplication tables

A concise and compact way of displaying the results of multiplying all the elements in a group together is a multiplication table. An example of such a table is shown in Table 2.4. All group elements are listed in the top row and the first column in the same order, beginning with the identity. The result of multiplying two elements appears at the intersection of the appropriate row and column. There is one thing that you should be aware of when using multiplication tables. Namely, pick a convention for the order of multiplication and stick to it. In this book, we will use the convention that elements across the top are applied first, then the element from the left-hand column: (left-hand side)(top). Also, a hint in constructing multiplication tables: each element appears in every row and column only once.

Exercise 2.6   Construct the multiplication table for the symmetry operations of the NH$_3$ molecule.

One final point regarding multiplication of group elements concerns the inverse of a product:

$$(AB...FG)^{-1} = G^{-1}F^{-1}...B^{-1}A^{-1} \tag{2.9}$$

This can be shown quite easily. First, write the product as

$$AB...FG = X \tag{2.10}$$

Then multiply on the right of Eq. 2.10 with G^{-1},

$$AB...FG(G^{-1}) = X(G^{-1}) \tag{2.11}$$

But the last term on the left-hand side of Eq. 2.11 is E. Continuing,

$$AB...F(F^{-1}) = X(G^{-1})(F^{-1}) \tag{2.12}$$

and so on. Finally,

$$E = X(G^{-1})(F^{-1})...(B^{-1})(A^{-1}) \tag{2.13}$$

and re-inserting Eq. 2.10 for X,

$$E = (AB...FG)(G^{-1})(F^{-1})...(B^{-1})(A^{-1}) \tag{2.14}$$

so that

$$(AB...FG)^{-1} = (G^{-1})(F^{-1})...(B^{-1})(A^{-1}) \tag{2.15}$$

The inverse of a product is the product of the inverses in reverse order.

It is a general result that the complete set of symmetry operations of an object form a group. Specifically,

(i) Closure: symmetry operations by definition transform the object into itself.
(ii) The identity operation, E, is always included in the set.
(iii) All symmetry operations have an inverse.
(iv) Associativity is axiomatic.

That the symmetry operations of an object form a group is an important statement because it means that we now have a mathematical framework for describing the symmetry of a crystal or a molecule. We will see throughout the rest of this book how this framework can be exploited to understand the electronic structure of materials.

2.8 Related group concepts

2.8.1 Subgroups

The group S is a subgroup of the group G if all the elements of S are also in G. It can be shown that if the order of the subgroup S is s and the order of G is h, then $h/s =$ integer. Note that for S to be a group, it must contain E.

2.8.2 Conjugate elements

Remember when we found the symmetry operations of the tetrahedron (Fig. 2.6)? You may recall that there were three identical σ_v operations but about different planes. The concept of conjugate elements will make the relationship among the different σ_v operations more concrete. Consider the elements $\{A, B, C, ...\}$ that form a group. A and B are *conjugate* elements if they are related by the transformation

$$A = X^{-1}BX \tag{2.16}$$

where X is *any member* of the group. For experts, this is a similarity transformation. One important point that should be stressed is that X can be *any* element in the group and it does not have to be *every* element in the group.

Some of the consequences of this definition are:

(i) Every element is conjugate with itself (let $X = E$):

$$E^{-1}AE = A \qquad (2.17)$$

$$E^{-1}BE = B \qquad (2.18)$$

(ii) If A is conjugate with B, then B is conjugate with A:

$$A = X^{-1}BX \qquad (2.19)$$

$$XAX^{-1} = B \qquad (2.20)$$

Or, simply swapping the two sides of Eq. 2.20,

$$B = XAX^{-1} \qquad (2.21)$$

But since X can be any member of the group, and X has an inverse in the group, we could equally well write $B = Y^{-1}AY$.

(iii) If A is conjugate to B and A is also conjugate to C, then B is conjugate to C. A is conjugate to B:

$$A = X^{-1}BX \qquad (2.22)$$

A is also conjugate to C:

$$A = Y^{-1}CY \qquad (2.23)$$

Therefore, setting Eqs. 2.22 and 2.23 equal:

$$X^{-1}BX = Y^{-1}CY \qquad (2.24)$$

Rearranging,

$$B = XY^{-1}CYX^{-1} \qquad (2.25)$$

Now let $Z = YX^{-1}$ so that $Z^{-1} = XY^{-1}$, which completes the proof.

2.8.3 Class

A set of elements of a group which are conjugate elements form a *class*. To find all the members in the same class as the element B, we apply the similarity transformation successively for each X in the group. This is indeed a tedious exercise! Some consequences are:

(i) Any member of one class will not be a member of any other class. This means that any group can be fully partitioned into classes.

(ii) E will always be in a class by itself.

(iii) E is the only class that is also a subgroup, since no other classes can contain E.

As an example, let us determine the classes of the group C_{2v}. There are no easy ways for determining conjugate elements except to multiply everything through. With experience, however, you will learn to recognize that elements in the same class (*i.e.*, conjugate elements) do the same sort of thing to objects. This means that we would not expect to find rotations and reflections in the same class.

The elements of C_{2v} are $\{E, C_2, \sigma_v, \sigma_v'\}$. What elements are conjugate to C_2? Using Table 2.4,

$$\sigma_v C_2 \sigma_v = \sigma_v \sigma_v' = C_2 \tag{2.26}$$

$$\sigma_v' C_2 \sigma_v' = \sigma_v' \sigma_v = C_2 \tag{2.27}$$

$$C_2 C_2 C_2 = C_2 \tag{2.28}$$

Therefore, C_2 is self-conjugate and is in a class by itself. You can work through the same procedure with σ_v and σ_v' and you will find that all elements are self-conjugate and are, hence, in their own classes. You can verify your partitioning of the group C_{2v} into four classes with the character table in Appendix 3.

2.8.4 Isomorphism

Two groups are *isomorphic* with each other if there is a one-to-one correspondence between the elements of the groups. Thus, the order of the groups must be the same and the group multiplication tables must be identical. This is an extremely powerful concept, as it will allow us to use abstract concepts to describe the symmetry properties of objects.

2.8.5 Homomorphism

Two groups are homomorphic with each other if there is a many-to-one correspondence between the elements of the groups. Thus, the order of the groups is different and the multiplication tables differ for the two groups.

2.9 Some special groups

We point out two special groups here that we will make reference to later.

2.9.1 Abelian groups

We have noted that, in general, we do not require the elements of a group to *commute*. That is, we do not require that $AB = BA$. The order of multiplication is important. However, for some groups, the order is not important because the elements commute. Such groups are called *Abelian*.

Exercise 2.7   Can you see that each element of an Abelian group must be in a class by itself?

2.9.2 Cyclic groups

Examination of the tables in Appendix 3 reveals some interesting properties of the groups C_n. Specifically, the group C_2 contains only E and C_2. Note that $C_2C_2 = E$. Every element in the group can be generated by repeated application of one element! These groups are called *cyclic*. Cyclic groups are also Abelian, which means that each element is in a class by itself. We will see later that the group of translation operations of a crystal is a cyclic group.

Problems

2.1   Draw the stereograms for the following symmetry operations: (a) E, i; (b) E, C_2; (c) E, C_2, C_4, C_4^3; (d) E, i, C_2, σ_h.

2.2   Find the point group for each of the collections of symmetry operations in Problem 2.1 above.

2.3   What symmetry operations (both point and nonpoint) do the figures in Fig. 2.12 exhibit?

2.4   What is the symmetry of a cube when a line is drawn across each of its faces as in Fig. 2.13? If alternate vertices of the marked cube are painted black, what is the symmetry? By symmetry we mean to what point group does the object belong?

(a)

(b)

(c)

(d)

(e)

Fig. 2.12.   From B. Grünbaum and G.C. Shephard, *Tilings and Patterns: An Introduction* (W.H. Freeman and Company, New York, 1989), pp. 216–217. Used with permission.

2.5   What point groups result on adding a center of symmetry to the point groups (a) C_1; (b) C_2; (c) C_3; (d) D_2; (e) C_{4v}?

2.6   What point groups result from the combination of two mirror planes at (a) 90° to each other; (b) 60°; (c) 45°; (d) 30°?

2.7   Determine the multiplication table for the point groups (a) C_3; (b) D_3; (c) C_{3v}; and (d) the permutation group of order 3. This last group contains as elements three objects labeled 1, 2, and 3. The

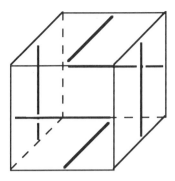

Fig. 2.13. Cube with line drawn across each face. See Problem 2.4.

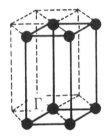

Fig. 2.14. The primitive hexagonal unit cell. The point at the center of the cell, Γ, is labeled. See Problem 2.10.

identity element is the objects in their numerological order (1)(2)(3). The operation of this group is permutation; that is, object 1 goes to the position of object 2 and *vice versa*. Object 3 remains where it is. It helps to write down all possible elements in this group first. There are six.

2.8 Prove that if A is conjugate to B and A is conjugate to C, then B is conjugate to C.

2.9 Of the point groups C_4, D_2, and C_{2h}, which are Abelian? Which are cyclic?

2.10 Shown in Fig. 2.14 is a primitive hexagonal unit cell. The point Γ is marked on the figure.

   (a) Find the symmetry operations at the Γ point. Verify that this set of operations forms a group. What are the symmetry elements associated with this point? Is this group Abelian? What is the group?

(b) Using stereographic projection, construct the multiplication table for the symmetry operations of this point.

(c) Partition the symmetry operations of the group into classes by finding the conjugate elements. Do the elements that belong to the same class do the same sorts of things to the object?

2.11 Consider groups of order $h = 1, 2, 3, 5$, and 7. Argue that these groups must be cyclic groups. Are these groups also Abelian? What can you say about the subgroups of these groups? Based on your observations of these groups, what can you say about the group of order $h = 47$?

3

Matrix representations and characters of finite groups

We have learned in the previous chapters about the necessity of group theory for understanding bonding in crystals (or at least in interpreting band structures), and we have already gone over symmetry operations and some formal concepts of group theory. In this chapter we will show that the symmetry operations of an object can be represented mathematically by a set of *square matrices*. The set of square matrices is homomorphic (many-to-one correspondence) to the set of symmetry operations. What is *unique* about the matrices in a representation is their character, or sum of the diagonal elements, to which we turn in the latter part of this chapter. First, however, let us investigate matrix representations for symmetry operations.

3.1 Matrix representations

One can show that the following set of 2×2 matrices form a group:

$$E = \begin{bmatrix} 1 & 0 \\ 0 & 1 \end{bmatrix}$$

$$A = \frac{1}{2} \begin{bmatrix} -1 & \sqrt{3} \\ -\sqrt{3} & -1 \end{bmatrix}$$

$$B = \frac{1}{2} \begin{bmatrix} -1 & -\sqrt{3} \\ \sqrt{3} & -1 \end{bmatrix}$$

$$C = \begin{bmatrix} -1 & 0 \\ 0 & 1 \end{bmatrix}$$

Table 3.1. *Multiplication table for the matrices* $\{E, A, B, C, D, F\}$

	E	A	B	C	D	F
E	E	A	B	C	D	F
A	A	B	E	F	C	D
B	B	E	A	D	F	C
C	C	D	F	E	A	B
D	D	F	C	B	E	A
F	F	C	D	A	B	E

$$D = \frac{1}{2} \begin{bmatrix} 1 & -\sqrt{3} \\ -\sqrt{3} & -1 \end{bmatrix}$$

$$F = \frac{1}{2} \begin{bmatrix} 1 & \sqrt{3} \\ \sqrt{3} & -1 \end{bmatrix}$$

The group operation is matrix multiplication. One can also show that these matrices have the multiplication table shown in Table 3.1.

Exercise 3.1   Show that the collection of matrices forms a group and that they have the multiplication table so indicated.

It turns out that this set of matrices is isomorphic with the group of symmetry operations C_{3v}. This is true because the two groups have the same number of elements and the same multiplication table.

Exercise 3.2   What is the correspondence between the matrices above and the symmetry operations of C_{3v}? That is, the matrix A corresponds to what operation, *etc.*?

Where did these matrices come from and how do we get more of them? The answer is simple: we looked at how the three cartesian coordinates x, y, z transform among themselves under all the symmetry operations of the group.

Let us show this in more detail. We will denote the coordinates of a right-handed coordinate system as x_i for $i = 1, 2, 3$, *i.e.*, $x_1 = x$, $x_2 = y$, etc. After operating on these coordinates with some symmetry operation we will obtain the coordinates x'_j. The relation between the two sets of coordinates is given by:

$$x'_i = \sum_j \Gamma_{ij} x_j \tag{3.1}$$

or, written out in matrix form:

$$
\begin{bmatrix} x_1' \\ x_2' \\ x_3' \end{bmatrix} = \begin{bmatrix} \Gamma_{11} & \Gamma_{12} & \Gamma_{13} \\ \Gamma_{21} & \Gamma_{22} & \Gamma_{23} \\ \Gamma_{31} & \Gamma_{32} & \Gamma_{33} \end{bmatrix} \begin{bmatrix} x_1 \\ x_2 \\ x_3 \end{bmatrix}
$$

Consider a simple application to the group C_{3v}. Recall that the elements of this group are $\{E, 2C_3, 3\sigma_v\}$. What effect does E have on the vector $\vec{x} = x_1\hat{x}_1 + x_2\hat{x}_2 + x_3\hat{x}_3$? The answer, of course, is that it leaves it in precisely the same position,

$$
\begin{bmatrix} x_1' \\ x_2' \\ x_3' \end{bmatrix} = \begin{bmatrix} 1 & 0 & 0 \\ 0 & 1 & 0 \\ 0 & 0 & 1 \end{bmatrix} \begin{bmatrix} x_1 \\ x_2 \\ x_3 \end{bmatrix}
$$

The unit matrix is thus the appropriate *matrix representation* for the group element E. What about C_3? C_3 rotates the vector $\vec{x}$ about the x_3 axis by $2\pi/3$. Recall that our convention states that the rotation is in the sense of a right-hand screw (this means counterclockwise if we look down the x_3 axis). Therefore,

$$
\begin{bmatrix} x_1' \\ x_2' \\ x_3' \end{bmatrix} = \frac{1}{2} \begin{bmatrix} -1 & \sqrt{3} & 0 \\ -\sqrt{3} & -1 & 0 \\ 0 & 0 & 2 \end{bmatrix}
$$

A set of general matrices for each type of symmetry operation is given in Fig. 3.1.

Note that nothing in the last column or along the bottom row of any of the matrices in Fig. 3.1 gets interchanged with any other column or row. Because x_3 never transforms into x_1 or x_2, but always keeps to itself, the extra dimension is actually redundant. Therefore, we could just as easily use the set of 2×2 matrices obtained by eliminating the last row and column from our 3×3 matrices. In that case, the matrix above for C_3 is identical with the matrix A at the beginning of this section. The collection of matrices $\{E, A, B, C, D, F\}$ form or *generate* a *representation* for the group of symmetry operations $\{E, 2C_3, 3\sigma_v\}$. We use the notation $\Gamma(A)$ to indicate a representation of the symmetry operation A. This notation says nothing about the dimensionality of the representation or the matrix elements, except that it is always understood that the matrices are square.

Exercise 3.3   Verify that the remaining matrices given at the beginning of this section correspond to the remaining symmetry operations of the group C_{3v}.

$$E = \begin{pmatrix} 1 & 0 & 0 \\ 0 & 1 & 0 \\ 0 & 0 & 1 \end{pmatrix} \qquad C_n = \begin{pmatrix} \cos\frac{2\pi}{n} & -\sin\frac{2\pi}{n} & 0 \\ \sin\frac{2\pi}{n} & \cos\frac{2\pi}{n} & 0 \\ 0 & 0 & 1 \end{pmatrix}$$

$$i = \begin{pmatrix} -1 & 0 & 0 \\ 0 & -1 & 0 \\ 0 & 0 & -1 \end{pmatrix} \qquad \sigma_h = \begin{pmatrix} 1 & 0 & 0 \\ 0 & 1 & 0 \\ 0 & 0 & -1 \end{pmatrix}$$

$$S_n = \begin{pmatrix} \cos\frac{2\pi}{n} & -\sin\frac{2\pi}{n} & 0 \\ \sin\frac{2\pi}{n} & \cos\frac{2\pi}{n} & 0 \\ 0 & 0 & -1 \end{pmatrix}$$

Fig. 3.1. Matrices representing general point symmetry operations.

Consider next the point group C_{4v}. The symmetry operations of this group are $\{E, 2C_4, C_2, 2\sigma_v, 2\sigma_d\}$. This group is very similar to the case of C_{3v} in that a set of 2×2 matrices suffices to represent the symmetry operations of the group. Again, the x_3 dimension is redundant because none of the C_{4v} symmetry operations interchange the x_3 coordinate with either x_1 or x_2.

Exercise 3.4   Work out the set of 2×2 matrices for the group C_{4v}.

There is nothing special about using the three coordinates x_1, x_2, and x_3 to obtain matrix representations. We can use any functions of coordinates to generate a matrix representation. Note that the magnitude of a vector $|\vec{r}| = \sqrt{x^2 + y^2 + z^2}$ will always transform into itself.

Exercise 3.5   Prove this last statement.

We could just as well use the numbers 1 or -1 to generate a representation. There is thus a very large number of representations that can be generated for each group of symmetry operations.

Some functions, however, are intrinsically more interesting than others, particularly when we study the electronic structure of materials. Graphical representations of the s, p_z, d_{xz}, and d_{z^2} atomic orbitals of the hydrogen atom are shown in Figs. 3.2, 3.3, 3.4, and 3.5, respectively. Since s atomic orbitals are spherically symmetric, they will generate representations like $r = \sqrt{x^2 + y^2 + z^2}$, or all 1×1-dimensional matrices containing 1's. By considering the functions x, y, z, we have already found out how p atomic

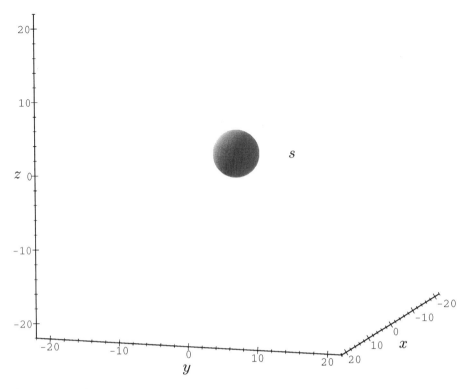

Fig. 3.2. Surface plot of the 1s atomic orbital of the hydrogen atom. The surface is constructed such that there is a 95% probability of finding the electron inside.

orbitals transform.† Another important set of functions is those that have the symmetry of d atomic orbitals: $z^2, x^2 - y^2, xy, xz, yz$.

Exercise 3.6   Based on how we found matrix representations for the three cartesian coordinates, can you figure out the matrix representations for the d atomic orbitals?

One could continue and work out the representations for the atomic orbitals of f, g, etc., symmetry.

Why are we so concerned about how the various atomic orbitals transform under the symmetry operations of point groups? When we come to solving for the electronic structure of a solid, we will discover that the wavefunction that describes the position of electrons can, in some cases, be written as a sum over these atomic orbitals. In other cases, we will have to resort to

† If this is not clear to you, we will review it in Ch. 6.

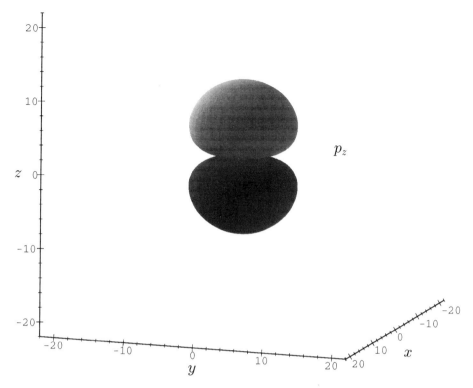

Fig. 3.3. Surface plot of the $2p_z$ atomic orbital of the hydrogen atom. The surface is constructed such that there is a 95% probability of finding the electron inside.

writing the wavefunction as a sum of plane waves. Finding the appropriate functions in which to cast these expansions is one of the major themes of electronic structure calculations.

Let us say a few brief words about equivalent representations. Suppose we have an l-dimensional matrix representation of a group of symmetry operations. Then an *equivalent representation* $\Gamma'(A)$ can be formed by the similarity transformation,

$$\Gamma'(A) = S^{-1}\Gamma(A)S \tag{3.2}$$

where S is a nonsingular $l \times l$ matrix (det $S \neq 0$) and A is any element of the group. The proof of this is given in any of the group theory texts in the Bibliography. The consequences of this theorem are important:

(i) one-dimensional representations are unique;

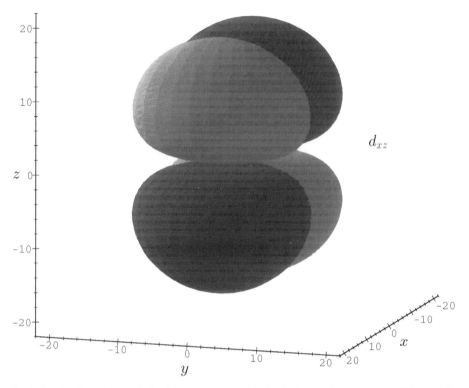

Fig. 3.4. Surface plot of the $3d_{xz}$ atomic orbital of the hydrogen atom. The surface is constructed such that there is a 95% probability of finding the electron inside.

(ii) higher-dimensional representations are not unique: one can find an infinite number of equivalent representations by repeated application of the similarity transformation of Eq. 3.2.

3.2 Irreducible representations

We have already seen in several examples above that we generated representations with a redundant dimension. Now we will see how systematically to find the smallest representation possible.

First, let us consider a set of $l \times l$ matrices that form a representation $\Gamma(R)$ of a group, where R is any element in the group. Do not confuse $\Gamma(R)$ with the individual components of the transformations that we discussed in the previous section. When we have need of the individual matrix elements, we will put subscripts on Γ. Suppose we can find a similarity transformation

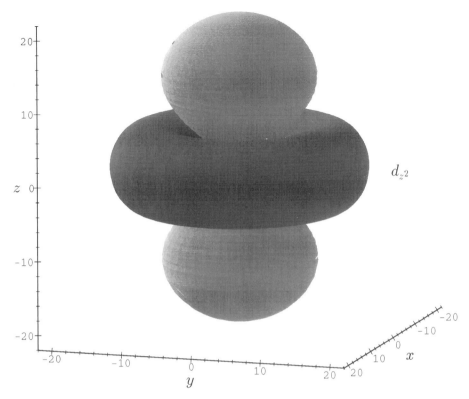

Fig. 3.5. The surface plot of the $3d_{z^2}$ atomic orbital of the hydrogen atom. Each surface is constructed such that there is a 95% probability of finding the electron inside.

that simultaneously transforms each $\Gamma(R)$ matrix into a block diagonal form,

$$\begin{bmatrix} \Gamma_1(R) & \mathbf{0} \\ \mathbf{0} & \Gamma_2(R) \end{bmatrix}$$

where $\Gamma_1(R)$ is an $l_1 \times l_1$ matrix and $\Gamma_2(R)$ is an $l_2 \times l_2$ matrix and $l_1 + l_2 = l$. The remaining blocks with zeros have the appropriate dimensions to assure that the matrix is square.

Each representation $\Gamma_1(R)$ and $\Gamma_2(R)$ is *irreducible* if neither one can be broken down into a smaller block diagonal form. $\Gamma_1(R)$ and $\Gamma_2(R)$ are important because *separately* they are representations of the group. This is clear, since if $\Gamma'(R)$ operates on some function, it will not mix up the two blocks formed by $\Gamma_1(R)$ and $\Gamma_2(R)$.

If a representation can be put into block diagonal form, then it is said to be *reducible*. This means that some similarity transformation exists that will put it into block diagonal form. In view of this, the 3×3 matrices we found in the previous section (Fig. 3.1) are *reducible*, while the 2×2 matrices $A, B, ...$ are *irreducible*.

A symbolic way of writing the decomposition of a reducible representation into its irreducible representations is by using the direct sum:

$$\Gamma'(R) \ = \ \Gamma_1(R) \oplus \Gamma_2(R) \tag{3.3}$$

This is *not* matrix addition in the usual sense. Beware! $\Gamma_1(R)$ and $\Gamma_2(R)$ may have different dimensions, as we have already seen above, and so we cannot add them in the traditional way.

Why are irreducible representations important? We will show shortly that any arbitrary representation of a group can be decomposed *uniquely* into irreducible representations of the group. In addition, the number of irreducible representations is small. Remember the problem that we encountered above in recognizing that the number of representations that we could find for a group was large? Irreducible representations will take care of this problem.

We will now state an important theorem about representations, whether they are reducible or irreducible. Recall that the character of a matrix is the sum of its diagonal elements.

Theorem 3.1 *The character of each matrix in a representation is unaltered by a similarity transformation.*

We demonstrate the proof to gain insight into why characters are important.

Proof The character can be written mathematically as:

$$\chi(R) \ = \ \sum_i (\Gamma(R))_{ii} \tag{3.4}$$

To prove this theorem, we need to show that the characters of $\Gamma'(R)$ and $\Gamma(R)$ are identical. From Eq. 3.2,

$$
\begin{aligned}
\chi(\Gamma'(R)) \ &= \ \chi(S^{-1}\Gamma(R)S) \\
&= \ \sum_i (S^{-1}\Gamma(R)S)_{ii} \\
&= \ \sum_i \sum_j \sum_k (S^{-1})_{ij}(\Gamma(R))_{jk}(S)_{ki} \\
&= \ \sum_i \sum_j \sum_k (S)_{ki}(S^{-1})_{ij}(\Gamma(R))_{jk}
\end{aligned}
$$

$$= \sum_j \sum_k (SS^{-1})_{kj} (\Gamma(R))_{jk}$$

$$= \sum_j \sum_k \delta_{kj} (\Gamma(R))_{jk}$$

$$= \sum_k (\Gamma(R))_{kk}$$

$$\chi(\Gamma'(R)) = \chi(\Gamma(R)) \tag{3.5}$$

□

This theorem is important because it brings out the importance of class. Elements in the same class are conjugate elements and are related by a similarity transformation. The characters of matrices that represent symmetry operations in the same class are identical! Therefore, we need not be concerned with every element in the group; only the classes are unique.

3.3 Great Orthogonality Theorem

The 'Great Orthogonality Theorem', which also goes by the name of the 'Grand Orthogonality Theorem' and the acronym GOT, is a basic theorem that relates the elements of the matrices that constitute the irreducible representations of a group. We will state this theorem without proof, although the reader may refer to the Bibliography for ideas.

GOT describes the orthogonality among different representations and the rows and columns within the matrices that constitute the different representations.

Theorem 3.2 *If Γ_i and Γ_j are two nonequivalent (i.e., cannot be related by a similarity transformation) irreducible unitary† representations, then*

$$\sum_{R=1}^{h} (\Gamma_i(R))_{mn}^* (\Gamma_j(R))_{op} = \frac{h}{l_i} \delta_{ij} \delta_{mo} \delta_{np} \tag{3.6}$$

where:

- *the sum runs over* all *the group elements (not just classes);*
- *each $\Gamma(R)$ is a matrix with elements (m, n) or (o, p) describing the effects of a symmetry operation on an object;*
- ** means take the complex conjugate;*
- *i and j are different irreducible representations generated, for example, by p orbitals or d orbitals;*

† A unitary matrix is a matrix for which the inverse is equal to the complex conjugate of the transpose. A unitary representation consists of all unitary matrices.

- h is the order of the group;
- l_i is the dimension of the ith irreducible representation;
- δ_{ij}, etc., are Kronecker delta functions; $\delta_{ij} = 1$ for $i = j$ and $\delta_{ij} = 0$ for $i \neq j$.

Briefly, what does this horrible-looking expression mean? A helpful physical picture is to think about taking the same matrix element from each representation and forming an h-dimensional vector out of them. That is, the number of *components* of each vector is equal to the order of the group. One vector can be constructed by taking the (m, n) element from *each* matrix in the ith representation for the group. Another vector can be constructed by taking the (o, p) element from *each* matrix in the jth representation for the group. GOT then tells us about the relationships of these vectors. Let us look specifically at three consequences.

　(i)　If $i \neq j$ (different irreducible representations),

$$\sum_{R=1}^{h} (\Gamma_i(R))_{mn}^* (\Gamma_j(R))_{mn} = 0 \qquad (3.7)$$

　　　This means that vectors from different representations are orthogonal.
　(ii)　If $m \neq o$ and/or $n \neq p$, but $i=j$,

$$\sum_{R=1}^{h} (\Gamma_i(R))_{mn}^* (\Gamma_i(R))_{op} = 0 \qquad (3.8)$$

　　　This means that vectors chosen from the *same* representation but from different elements are orthogonal.
　(iii)　Let $m=o$, $n=p$, and $i=j$:

$$\sum_{R=1}^{h} (\Gamma_i(R))_{mn}^* ((\Gamma_i(R))_{mn} = \frac{h}{l_i} \qquad (3.9)$$

　　　This means that the length of the vector is equal to the square root of h/l_i.

Essentially, then, GOT tells us about restrictions placed on the elements that comprise the irreducible representations. Let us look at an example to try to clarify GOT.

The matrices that are listed at the beginning of this chapter are good representations of the point group C_{3v}, provided we get rid of the redundant dimension and use only 2×2-dimensional matrices. Using the coordinate x_3, we can construct a second representation for the group C_{3v}. This amounts

Table 3.2. *The three irreducible representations of the point group C_{3v}*

	Γ_1	Γ_2	Γ_3
E	1	1	$\begin{bmatrix} 1 & 0 \\ 0 & 1 \end{bmatrix}$
C_3	1	1	$\frac{1}{2}\begin{bmatrix} -1 & \sqrt{3} \\ -\sqrt{3} & -1 \end{bmatrix}$
C_3^2	1	1	$\frac{1}{2}\begin{bmatrix} -1 & -\sqrt{3} \\ \sqrt{3} & -1 \end{bmatrix}$
σ_v^1	1	-1	$\begin{bmatrix} -1 & 0 \\ 0 & 1 \end{bmatrix}$
σ_v^2	1	-1	$\frac{1}{2}\begin{bmatrix} 1 & -\sqrt{3} \\ -\sqrt{3} & -1 \end{bmatrix}$
σ_v^3	1	-1	$\frac{1}{2}\begin{bmatrix} 1 & \sqrt{3} \\ \sqrt{3} & -1 \end{bmatrix}$

to looking at the elements in the redundant dimension from the set of 3×3 matrices. This representation is denoted by Γ_1 and it has the form

$$\begin{array}{ccccccc} & E & C_3 & C_3^2 & \sigma_v^1 & \sigma_v^2 & \sigma_v^3 \\ \Gamma_1 & 1 & 1 & 1 & 1 & 1 & 1 \end{array}$$

That is, to each symmetry operation there corresponds the 1×1 matrix containing $+1$.

We can generate a third representation using combinations of 1 and -1; it is given by the following and is denoted Γ_2:

$$\begin{array}{ccccccc} & E & C_3 & C_3^2 & \sigma_v^1 & \sigma_v^2 & \sigma_v^3 \\ \Gamma_2 & 1 & 1 & 1 & -1 & -1 & -1 \end{array}$$

Let us call the group of matrices generated at the beginning of the chapter Γ_3. Clearly Γ_1 and Γ_2 are irreducible, since they are one-dimensional. Maybe you can show that Γ_3 is also irreducible? The irreducible representations for the point group C_{3v} are collected in Table 3.2.

Choose $\Gamma_i = \Gamma_j = \Gamma_1$. Then applying GOT,

$$\begin{aligned} \sum_{R=1}^{6} \Gamma_1(R)^* \Gamma_1(R) &= 1^2 + 1^2 + 1^2 + 1^2 + 1^2 + 1^2 \\ &= 6 \\ &= \frac{h}{l_1} \end{aligned} \tag{3.10}$$

Similar results are found if we use $\Gamma_i = \Gamma_j = \Gamma_2$. However, if we let

$\Gamma_i = \Gamma_j = \Gamma_3$ and choose $m = o = 1$ and $n = p = 2$, then we find

$$
\begin{aligned}
\sum_{R=1}^{6} (\Gamma_3(R))_{12}^* (\Gamma_3(R))_{12} &= 0 + \frac{3}{4} + \frac{3}{4} + 0 + \frac{3}{4} + \frac{3}{4} \\
&= 3 \\
&= \frac{h}{l_2}
\end{aligned}
\tag{3.11}
$$

Note that Γ_3 is a two-dimensional representation.

Exercise 3.7   Work out as many more examples of GOT for the matrix representations of C_{3v} or C_{4v} as you can.

3.4 Properties of characters of irreducible representations

So far in this chapter we have introduced the idea that matrices can be used to represent the action of symmetry operations on the three cartesian coordinates. We also presented the 'Great Orthogonality Theorem' and showed a few of its implications for the irreducible representations of a point group. We now turn to examining some of the consequences of GOT for the *characters* of irreducible representations.

Some of these consequences are:

(i) The characters of matrices in the same class are identical.

   We have already shown that the characters are unchanged by a similarity transformation, Eq. 3.5. Now identify S, the nonsingular matrix of Eq. 3.2, with a matrix representation of a symmetry operation and the consequence is clear.

(ii) When summed over all the symmetry operations, R, of the group, the characters of irreducible representations are orthogonal and normalized to the order of the group, h.

$$
\sum_{R=1}^{h} \chi_i(R)^* \chi_j(R) = h \delta_{ij}
\tag{3.12}
$$

This result follows directly from GOT. For characters, take $m = n$ and $o = p$,

$$
\sum_{R=1}^{h} (\Gamma_i(R))_{mm}^* (\Gamma_j(R))_{pp} = \frac{h}{l_i} \delta_{ij} \delta_{mp} \delta_{mp}
\tag{3.13}
$$

and sum over m and p to obtain the trace,

$$\sum_m \sum_p \sum_{R=1}^h (\Gamma_i(R))^*_{mm} (\Gamma_j(R))_{pp} = \frac{h}{l_i} \delta_{ij} \sum_m \sum_p \delta_{mp} \delta_{mp}$$

$$\sum_{R=1}^h \chi_i(R)^* \chi_j(R) = \frac{h}{l_i} \delta_{ij} l_i$$

$$\sum_{R=1}^h \chi_i(R)^* \chi_j(R) = h \delta_{ij} \tag{3.14}$$

There are two implications of this result:

(a) If $i = j$, then

$$\sum_{R=1}^h \chi_i(R)^* \chi_i(R) = h \tag{3.15}$$

Although it may not be obvious, Eq. 3.15 is a necessary and sufficient condition for a representation to be irreducible.

Exercise 3.8   Verify this last statement for a few examples.

(b) If $i \neq j$, then

$$\sum_{R=1}^h \chi_i(R)^* \chi_j(R) = 0 \tag{3.16}$$

and the two different representations are orthogonal. We note in passing that the sums are over *all* symmetry operations, not just classes. However, we can rewrite these sums to be over classes.

(iii) A necessary and sufficient condition for two representations to be equivalent is that the characters of the representations must be equal.

Recall that equivalent representations are related by a similarity transformation,

$$\Gamma'(A) = S^{-1} \Gamma(A) S \tag{3.17}$$

where S is nonsingular. Sometimes it is not at all evident how we find this nonsingular matrix S to show that two representations are equivalent. GOT makes it unnecessary to find S. All we have to do is look at the characters, which are trivial to find once we have a representation.

Exercise 3.9   The characters of the three representations of the group C_{3v} in Table 3.2 are not equal. Are the representations equivalent?

(iv) If l_i is the dimension of the ith irreducible representation of a group, then

$$\sum_i l_i^2 = h \tag{3.18}$$

where the sum runs over all irreducible representations.

Exercise 3.10   From Table 3.2, $(1)^2 + (1)^2 + (2)^2 = 6$. Are there any more irreducible representations?

One important thing to recognize is that

$$l_i = \chi_i(E) \tag{3.19}$$

or, the dimension of an irreducible representation is the character of the identity. We can rewrite Eq. 3.18 as

$$\sum_i [\chi_i(E)]^2 = h \tag{3.20}$$

Eq. 3.18 puts constraints on the dimensionalities of the irreducible representations.

Exercise 3.11   Can you show why the dimension of an irreducible representation must be equal to the character of the identity?

(v) The number of *inequivalent* irreducible representations equals the number of classes of the group.

Remember that inequivalent irreducible representations cannot be related by a similarity transformation. We state this without proof and refer the interested reader to any of the group theory texts in the Bibliography for details.

As an example, refer once again to the three representations generated for C_{3v} in Table 3.2. Given the symmetry operations in a group, it is a tedious but straightforward task to find the classes. Then we know the number of irreducible representations. For the group C_{3v}, we know there are three classes, so there must be three irreducible representations and, hence, all of them are listed.

3.5 Characters of some point groups

Using the consequences of GOT that we have just stated, we now proceed to determine the characters for some groups.

(i) Groups with two symmetry operations, $\{E, A\}$. We must have two classes because E is always in a class by itself. There must be two irreducible representations. The one irreducible representation that all groups possess is the 'totally symmetric' representation that consists of all '1"s. For groups containing two elements, the other irreducible representation must be one-dimensional to satisfy

$$\sum_i [\chi_i(E)]^2 = h \qquad (3.21)$$

$$(1)^2 + (1)^2 = 2 \qquad (3.22)$$

Finally, from

$$\sum_{R=1}^{h} \chi_i(R)^* \chi_j(R) = h\delta_{ij} \qquad (3.23)$$

we can construct the remaining unknown character. Clearly this must be -1.

There are three crystallographic point groups with $h=2$; these are $C_{1h}(E, \sigma_h)$, $S_2(E, i)$, and $C_2(E, C_2)$. All these groups have the same characters and the same representations in spite of the fact that they contain different elements. They are therefore isomorphic.

We can neatly quantify the information about symmetry operations of a group, the characters, and the irreducible representations in a *character table*. The character table appropriate for groups containing two symmetry operations is shown in Fig. 3.6. We will have more to say about character tables shortly.

(ii) Noncyclic groups of order four. Remember that noncyclic means that the group cannot be generated by repeated application of one element. These groups have four classes and so they must have four irreducible representations.

Exercise 3.12   Can you prove that a noncyclic group of order four must have four classes?

Remember that we always have the totally symmetric irreducible representation of all $+1$'s. It remains to find the other three irreducible representations. For these groups, what must be the dimensions of the other irreducible representations? They must satisfy:

$$\sum_i [\chi_i(E)]^2 = h = 4 \qquad (3.24)$$

$$(1)^2 + [\chi_2(E)]^2 + [\chi_3(E)]^2 + [\chi_4(E)]^2 = 4 \qquad (3.25)$$

(a)

C_{1h}				E	σ_h
		S_2		E	i
			C_2	E	C_2
Γ_1	x, y	z^2, etc., xy, etc.	z	1	1
Γ_2	z	x, y, z	x, y	1	-1

(b)

D_2					E	C_2^z	C_2^y	C_2^x
		C_{2v}			E	C_2	σ'_v	σ_v
				C_2	E	C_2	i	σ_h
Γ_1	x^2, y^2, z^2	x^2, y^2, z^2	x^2, y^2, z^2, xy		1	1	1	1
Γ_2	y, xz, x^2y	x, xz	yz, xz		1	-1	1	-1
Γ_3	z, xy, y^2z	xy, xyz	z, y^2z		1	1	-1	-1
Γ_4	z, yz, xy^2	y, yz	y, x		1	-1	-1	1

Fig. 3.6.   Character tables for some groups of order (a) two and (b) four.

The only solution is $\chi_2(E) = \chi_3(E) = \chi_4(E) = 1$. So the other three irreducible representations are also one-dimensional.
   Using the relationship

$$\sum_R \chi_i(R)^* \chi_j(R) = h\delta_{ij} \tag{3.26}$$

we can find that the other characters (*i.e.*, those not associated with the element E) must be $+1, -1, -1$ with their arrangement under the various symmetry operations consistent with this equation.
   There are three crystallographic noncyclic point groups that fall into this category; these are D_2, C_{2v}, and C_{2h}. Again, because the characters are the same with the same representations, these groups are isomorphic. See Fig. 3.6.
   One very important thing that can be ascertained immediately from a character table is how functions transform under the various symmetry operations. This is of the utmost importance, since we want to know how the various atomic orbitals are transformed by symmetry operations. Let us consider one case in detail so that this is clear.

From Fig. 3.6 we have the character table for the point group C_{1h}. This group contains the symmetry operations $\{E, \sigma_h\}$. The two irreducible representations are denoted Γ_1 and Γ_2. The extra column in the character table contains cartesian coordinates. Under the operation E, all of the cartesian coordinates will go into themselves. This is why the character of E is always a positive integer. In the C_{1h} case, x and y are grouped together into the first irreducible representation Γ_1, while z is by itself in the second irreducible representation Γ_2. Clearly, however, the characters associated with all three coordinates for the identity operation are equal. What about under the operation of σ_h? x and y remain unchanged, but z gets reflected through a horizontal mirror plane, thus changing its sign. But that is exactly what is shown in the character table! The character for x and y is the same and we say that, separately, they *transform as, or generate, the Γ_1 irreducible representation.* z transforms into minus itself under σ_h and, hence, its character is '−1'. The same thing holds true for the other two-dimensional groups that are isomorphic to C_{1h}.

We can gain important insight into how *products* of functions transform under symmetry operations by examining character tables. Consider the noncyclic groups of order 4, the character table of which is shown in Fig. 3.6. Take the point group C_{2h} as an example. z transforms as Γ_3, while y transforms as Γ_4. How might we expect their product to transform? The product transforms as the product of the characters! Let us verify this. First, under E, both y and z transform into themselves, so their product must also transform into itself. Hence, the character associated with the product is $+1$. Next, under C_2, y goes into minus itself, while z transforms into itself. The product thus transforms into minus itself. Under i, both functions transform into minus themselves, which means that the product must transform into plus itself. Finally, under σ_h, y goes into itself, while z transforms into minus itself, so that the product transforms into minus itself. The characters that we find for the product are thus:

$$\begin{array}{cccc} E & C_2 & i & \sigma_h \\ 1 & -1 & 1 & -1 \end{array}$$

But this is precisely the Γ_2 irreducible representation! In fact, the product of any one of the infinite number of functions that transform as Γ_4 multiplied by a function that transforms as Γ_3 will transform as Γ_2 because $\Gamma_2 = \Gamma_3 \times \Gamma_4$. It may be stated generally, then, that for one-dimensional representations, the characters express how the functions transform into themselves.

(iii) More complicated groups. We can find the characters of groups more complicated than those just discussed above. All characters are found in precisely the same way. The one aspect that requires special care is determining how various functions transform when the irreducible representation is two- or higher-dimensional. In this case, coordinates may be interchanged under action of a symmetry operation, *e.g.*, x and y transform into y and $-x$, respectively, under a C_4 rotation. We would say, then, that x and y *together* generate one of the irreducible representations.

3.6 Character tables

All character tables can be constructed by following the steps listed below:

(i) All groups possess the 'totally symmetric' irreducible representation that consists of all $+1$'s. This irreducible representation is listed as the topmost row in a character table.

(ii) The total number of irreducible representations is equal to the number of classes.

(iii) The dimensionality of the irreducible representations must satisfy

$$\sum_i [\chi_i(E)]^2 = h \tag{3.27}$$

(iv) The characters must be normalized:

$$\sum_{R=1}^{h} \chi_i^*(R)\chi_i(R) = h \tag{3.28}$$

(v) The characters must be orthogonal:

$$\sum_{R=1}^{h} \chi_i^*(R)\chi_j(R) = h\delta_{ij} \tag{3.29}$$

If you follow these rules, in this suggested sequence, you cannot go wrong in constructing a character table. That does not mean, of course, that construction of a character table is not a tedious task!

Let us briefly discuss the matter of notation. We will use two sets of notation for labeling irreducible representations. The notation that we have so far employed, $\Gamma_1, \Gamma_2, etc.$, is due to Bethe and it is arbitrary except that Γ_1 is *always* the totally symmetric representation. The other notation that we will use is that of Mulliken. The following subscripts are used in the Mulliken notation:

$$\left.\begin{array}{c} A \\ B \end{array}\right\} \; \textit{one-dimensional irreducible representations}$$

$$E \; \} \; \textit{two-dimensional irreducible representations}$$

$$T \; \} \; \textit{three-dimensional irreducible representations}$$

Exercise 3.13   Why are there no four-dimensional irreducible representations? (The icosahedral group is an exception.)

To complete the Mulliken notation, we must examine the symmetry operations contained in the group. If the irreducible representation is one-dimensional and *symmetric* under rotation about the principal axis, then the representation is given the symbol A. If it is one-dimensional and *antisymmetric*, it is given the symbol B. By way of example, consider the group C_{4v}. (See Appendix 3.) This group has a four-fold rotation axis. The characters of the two 'A' representations are $+1$ under C_4, while the characters of the two 'B' representations are -1 under C_4. Note, however, that the characters of the C_2 operation are all $+1$. The appropriate notation comes from the highest rotation about the principal axis.

If the group contains a C_2 rotation perpendicular to the principal axis, then a subscript '1' or '2' is appended to the irreducible representation symbol. If the perpendicular C_2 rotation is symmetric, we add a subscript '1'. If the rotation is antisymmetric, we add a subscript '2'. If there is *no* perpendicular C_2, then we examine symmetry with respect to a σ_v operation. The group C_{4v} is a good example of this use of notation.

If i is a symmetry operation of the group, then the functions of those irreducible representations that transform into $+1$ times themselves are given the subscript 'g', which is from the German 'gerade' for even. The functions of those irreducible representations that transform into -1 times themselves are given the subscript 'u', or 'ungerade' for odd. A simple example is provided by the point group S_2 in Table 3.3.

Finally, if the group contains no inversion operation, then we determine the symmetry with respect to the σ_h operation. If the irreducible representation is symmetric under σ_h, then it is given a superscript prime ($'$). If it is antisymmetric under σ_h, it is given a superscript double prime ($''$).

A generic character table has the general structure shown in Fig. 3.7. The rightmost side of a character table is often divided into two sections. In

Table 3.3. *Character table for the point group* S_2

Mulliken	Bethe	E	i
A_g	Γ_1	1	1
A_u	Γ_2	1	-1

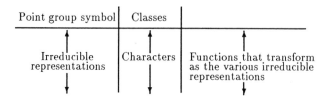

Fig. 3.7. The general structure of a character table.

the first subsection, we will always find the six symbols x, y, z, R_x, R_y, R_z. The first three are just the ordinary cartesian coordinates, while the R's stand for rotations about the subscripted axis. In the second subsection are listed all the square and binary products of coordinates according to their transformation properties.

3.7 Reduction of a reducible representation

Given some matrix representation, how do we know whether it is reducible or irreducible? What we need to do is find some matrix that puts Γ into block diagonal form. It is unlikely that this is going to be an efficient way. There are a number of other choices, however.

(i) The first thing we should do is to check the dimensionality of the representation. Does it fit in with the class structure of the group? Recall the example of the noncyclic groups of order 4 above.

(ii) Secondly, we can compare the characters of our given representation with those in a character table. If they are identical, then clearly it must be true that the representation is irreducible, since all the representations in character tables are irreducible.

(iii) If they are not identical, then the representation must be reducible. What is also true is that this reducible representation is comprised of irreducible representations. We can find out how many times our reducible representation contains each irreducible representation of the group:

$$\chi(R) = \sum_i n_i \chi_i(R) \tag{3.30}$$

where $\chi(R)$ is the character of the reducible representation, $\chi_i(R)$ is the character of the ith irreducible representation, and n_i is the number of times the ith irreducible representation appears in the reducible representation. Multiplying by a character within the jth irreducible representation and summing over all symmetry operations in the group, we have

$$\sum_{R=1}^{h} \chi_j^*(R)\chi(R) = \sum_{i,R} n_i \chi_j^*(R)\chi_i(R) \tag{3.31}$$

Now, one consequence of GOT is that

$$\sum_{R=1}^{h} \chi_j^*(R)\chi_i(R) = h\delta_{ij} \tag{3.32}$$

so that Eq. 3.31 becomes

$$\sum_{R=1}^{h} \chi_j^*(R)\chi(R) = hn_j \tag{3.33}$$

Therefore,

$$n_j = \frac{1}{h} \sum_{R=1}^{h} \chi_j^*(R)\chi(R) \tag{3.34}$$

Or, n_j is the number of times the reducible representation contains the jth irreducible representation.

To demonstrate reduction of a reducible representation, Γ_{red}, consider the group C_{3v}, which has elements $\{E, 2C_3, 3\sigma_v\}$, and the characters comprising a given representation $\chi(R) = 5;\ 2;\ -1$. Using the character table for the point group C_{3v} which can be found in Appendix 3, we find

$$n_{A_1} = \frac{1}{6}[1(1)(5) + 2(1)(2) + 3(1)(-1)] = 1$$

$$n_{A_2} = \frac{1}{6}[1(1)(5) + 2(1)(2) + 3(-1)(-1)] = 2$$

$$n_E = \frac{1}{6}[1(2)(5) + 2(-1)(2) + 3(0)(-1)] = 1 \tag{3.35}$$

Therefore,

$$\Gamma_{\text{red}} = A_1 + 2A_2 + E \tag{3.36}$$

There are a few things of which to be aware. First, do not forget that you are summing over *symmetry operations* and not just *characters*. This is the reason for the factors of 2 and 3 in the second and third terms. Also, do not forget to divide by the order of the group. Your answer must always be an integer and the decomposition is unique!

Problems

3.1 Find a representation of the point group D_2 by considering how the functions x, y, and z transform under the symmetry operations of the group. How many equivalent representations have you found, if any?

3.2 For the E representation of the point group C_{4v}, write a representation using the functions (x, y) *both* as basis functions. Repeat for $(x + \imath y)$ and $(x - \imath y)$ as basis functions. Can you find a similarity transformation between the two sets of basis functions? What is the physical significance of this transformation?

3.3 For the hexagonal unit cell of Problem 2.10, find the matrices for *all* the symmetry operations acting on the three cartesian coordinates (x, y, z).

3.4 Write out a representation table for the point group C_{2h} as has been done in Section 3.3 for the group C_{3v}. Check that the characters of representations in the same class are equal. Check the GOT for some of the representations.

3.5 Work out the character table for the point group C_{4v}.

3.6 Obtain the character tables for the permutation groups P_2 and P_3. See Ch. 2, Problems. Note that these groups are isomorphic to C_2 and C_{3v}.

3.7 The character tables list basis functions that are single cartesian coordinates, x_i, or products of coordinates, $x_i x_j$. For a point group of your choosing, how would you determine the irreducible representations generated by functions of the type $x_i x_j x_k$?

3.8 Show that under the application of a general point symmetry operation, the vector $|\vec{r}\,| = \sqrt{x^2 + y^2 + z^2}$ is transformed into itself. Note that such a transformation is important when we consider which function (or functions) generate the totally symmetric irreducible representation and when we consider how wavefunctions transform under the action of point symmetry operations (*i.e.*, since $|\vec{r}\,|$ is unchanged, we need only concern ourselves with how the angular parts of the wavefunction transform).

3.9   Using GOT, show that

$$\sum_{R=1}^{h} \chi_i(R) = 0 \qquad (P3.1)$$

3.10   Write the matrices describing the effect on a point (x, y, z) of reflections
       in vertical planes which lie halfway between the xz and yz planes. By
       matrix methods determine what operations result when each of these
       reflections is followed by a reflection in the xy plane.

4

Review of the mathematics of quantum mechanics

In this chapter we will introduce and review some of the mathematics and notation that will make using quantum mechanics to study electronic structure simpler.

The single most important function that quantum mechanics tells us about is the wavefunction, which is generally given by the Greek letter psi, Ψ. Further, it is customary to let upper case Ψ refer to a wavefunction that depends on both position, $\vec{r}$, and time, t. A lower case ψ usually refers to a time-independent wavefunction. Ψ is important because its square $|\Psi|^2$ tells us about the probability of finding an electron in a region of space $d\vec{r}$ about $\vec{r}$ at time t. In the language of electronic structure, this probability is just the charge density and that tells us where bonds are in molecules and solids. The equation that the wavefunction must satisfy is the Schrödinger equation; it is impossible to solve exactly in all but a handful of simple cases. We will return to a discussion of the Schrödinger equation in the next chapter.

4.1 Bra-ket notation and the scalar product

Dirac proposed a condensed notation for representing possible *states* of a quantum mechanical system. It is a postulate of quantum mechanics that the *state* of a particle is defined, at a given instant, by a wavefunction $\psi(\vec{r})$. We will ignore time-dependent wavefunctions for the time being. However, $\psi(\vec{r})$ can be represented by several distinct choices of components or coordinates.

Consider the situation in ordinary three-dimensional space. The position of a point in space can be given by a collection of three numbers, which are the coordinates with respect to a predetermined set of axes. If we make another choice of axes, for example, by rotation of the original set, then there is another set of coordinates that now corresponds to the point. But the position in space hasn't actually changed! The concepts of vector algebra allow us to avoid being tied to a specific system of coordinate axes.

In quantum mechanics we use a similar approach. Each quantum state of a particle will be characterized by a *state vector* which belongs to an abstract space called the *state space* of the particle. Technically, this abstract space is known as a Hilbert space. Just as we define all the rules for relating vectors, finding their lengths, *etc.*, in real space, we will do the same thing in state space for quantum states.

Any element, or vector, of state space is called a 'ket' and is represented by the notation $|\rangle$. In particular, we can associate every wavefunction $\psi(\vec{r})$ with a ket $|\psi\rangle$. Notice the fact that the $\vec{r}$-dependence no longer appears in the ket notation. $\psi(\vec{r})$ can be interpreted as the set of components of $|\psi\rangle$ in a particular basis, of which $\vec{r}$ is the index.

For each pair of kets $|\phi\rangle$ and $|\psi\rangle$ we can define a scalar, or dot, product $(|\phi\rangle, |\psi\rangle)$. The scalar product is a complex number that satisfies the following properties:

$$(|\phi\rangle, |\psi\rangle) = (|\psi\rangle, |\phi\rangle)^* \tag{4.1}$$

$$(|\phi\rangle, a_1|\psi_1\rangle + a_2|\psi_2\rangle) = a_1(|\phi\rangle, |\psi_1\rangle) + a_2(|\phi\rangle, |\psi_2\rangle) \tag{4.2}$$

$$(a_1|\phi_1\rangle + a_2|\phi_2\rangle, |\psi\rangle) = a_1^*(|\phi_1\rangle, |\psi\rangle) + a_2^*(|\phi_2\rangle, |\psi\rangle) \tag{4.3}$$

Note particularly in Eq. 4.3 that the constants a_1 and a_2 must be replaced by their complex conjugates a_1^* and a_2^*, respectively. We can simplify these expressions, actually, by noting that to every ket $|\psi\rangle$ there corresponds a 'bra' denoted by $\langle\psi|$. The origin of the terminology is the word 'bracket' for the symbol $\langle|\rangle$. Hence the name 'bra' for the left-hand symbol and 'ket' for the right-hand symbol. This correspondence is not actually as innocuous as it appears, but to delve further into its meaning would require a more advanced and subtle treatment than we want to provide.

In Dirac notation, then, the properties of the scalar product may be expressed succinctly as:

$$\langle\phi|\psi\rangle = \langle\psi|\phi\rangle^* \tag{4.4}$$

$$\langle\phi|a_1\psi_1 + a_2\psi_2\rangle = a_1\langle\phi|\psi_1\rangle + a_2\langle\phi|\psi_2\rangle \tag{4.5}$$

$$\langle a_1\phi_1 + a_2\phi_2|\psi\rangle = a_1^*\langle\phi_1|\psi\rangle + a_2^*\langle\phi_2|\psi\rangle \tag{4.6}$$

and $\langle\psi|\psi\rangle$ is real and positive and it can be zero if and only if $|\psi\rangle \equiv 0$.

Exercise 4.1   For a wavefunction $\psi(\vec{r}) = a_1\psi_1 + a_2\psi_2$, where a_1 and a_2 are arbitrary complex constants, can you show that $\langle\psi|\psi\rangle$ is real and greater than zero?

Positive definiteness is reassuring, since we want the scalar product to correspond to a charge density which must be real and positive! Making contact with something that we are more familiar with, we note that

$$\langle\psi|\psi\rangle \equiv \int \psi^*(\vec{r})\,\psi(\vec{r})\,d\vec{r} \tag{4.7}$$

where the integration is performed over all three-dimensional space. Since the probability density corresponds to a physical quantity, the integral in Eq. 4.7 must converge. But $d\vec{r} = r^2 dr \sin\theta d\theta d\phi$, which has units of volume; for the integral to be finite as $r \to \infty$, the wavefunction must decrease *faster* than $1/r$. Mathematically, we say that the wavefunction $\psi(\vec{r})$ must be *square integrable*.

Exercise 4.2   Show that $\psi(\vec{r}) = 1/r$ is not a square integrable function.

Besides being square integrable, our quantum mechanical wavefunction must be well behaved. This means that $\psi(\vec{r})$ must be everywhere defined, continuous, and infinitely differentiable.

Exercise 4.3   Using dimensional analysis, can you show why $|\psi(\vec{r})|^2$ is a probability *density*?

Finally, since the probability of locating an electron somewhere in space must be finite, we demand that our wavefunctions be *normalized*. That is,

$$\langle\psi|\psi\rangle = 1 \tag{4.8}$$

Typically $|\psi\rangle$ will contain a constant factor that is used to satisfy Eq. 4.8.

4.2 Expansion of basis states

As we know from dealing with vectors in three-dimensional space, we can use any set of orthonormal basis vectors to express any vector. An orthonormal basis set is both *orthogonal* and *normalized* to unit length,

$$\hat{x}_i \cdot \hat{x}_j = \delta_{ij} \tag{4.9}$$

where $i = 1, 2, 3$, and $\hat{x}_i$ is a unit vector and one of our ordinary cartesian axes.

We can do precisely the same thing with quantum mechanical state vectors. Suppose we have an orthonormal set of basis vectors $\{|\phi_i\rangle\}$ that is also complete. By complete we mean that the basis vectors are linearly independent and 'span' the space. That is, any vector can be expressed in terms of this complete set of basis vectors. Proof that a set of basis vectors is complete is something that we will always leave to the mathematicians. We can express the state vector $|\psi\rangle$ in the basis $\{|\phi_i\rangle\}$ as

$$|\psi\rangle = \sum_i c_i|\phi_i\rangle \tag{4.10}$$

where the sum runs over all the basis states and the c_i's are the expansion coefficients. But where do we get the expansion coefficients from? Multiplying both sides of Eq. 4.10 by the bra $\langle\phi_j|$, which is also a member of the orthonormal basis set,

$$\langle\phi_j|\psi\rangle = \sum_i c_i\langle\phi_j|\phi_i\rangle \tag{4.11}$$

and using $\langle\phi_i|\phi_j\rangle = \delta_{ij}$, we obtain

$$c_j = \langle\phi_j|\psi\rangle \tag{4.12}$$

If the set $\{|\phi_i\rangle\}$ truly forms a basis, then every wavefunction $|\psi\rangle$ can be expanded uniquely in terms of the basis functions. That is, there is only one set of coefficients c_i for each wavefunction.

Recall that we associate a wavefunction $\psi(\vec{r})$ with a ket $|\psi\rangle$. Now, suppose that we are searching for the wavefunction of an electron in anything as simple as a diatomic molecule to a complex crystal like the high-T_c superconductor $YBa_2Cu_3O_{7-\delta}$. One technique that we will make repeated use of is to express the wavefunction of the molecule or crystal as a sum over wavefunctions of the individual constituent *atoms*. This is known as the tight-binding approximation because it assumes that each electron in the more complex system is closely associated with the atom that contributes it. Plane waves, functions of the form $(1/2\pi\hbar)\exp(\imath px/\hbar)$, are also frequently used as basis functions.

Exercise 4.4   Can you see any potential difficulties in using plane waves as basis functions?

If we insert Eq. 4.12 into Eq. 4.10, we obtain

$$|\psi\rangle = \sum_i \langle\phi_i|\psi\rangle|\phi_i\rangle \tag{4.13}$$

which, on rearranging, yields

$$|\psi\rangle = \sum_i |\phi_i\rangle\langle\phi_i|\psi\rangle$$

$$= \left(\sum_i |\phi_i\rangle\langle\phi_i|\right)|\psi\rangle \qquad (4.14)$$

But since $|\psi\rangle$ appears on both sides of Eq. 4.14, what appears in parentheses is the identity

$$\sum_i |\phi_i\rangle\langle\phi_i| = \mathbf{1} \qquad (4.15)$$

The symbol on the right-hand side of Eq. 4.15, $\mathbf{1}$, can be thought of as the unit matrix, as will become clear after the following section. Eq. 4.15 is also said to express closure of the basis set $\{|\phi_i\rangle\}$.

Exercise 4.5   Using the expansion for a wavefunction, Eq. 4.10, show that the properties of the scalar product can be re-expressed in terms of the components of the basis set. Be sure not to confuse the basis vectors with the wavefunctions! Your results characterize the state of a quantum mechanical particle just as well as the wavefunction $\psi(\vec{r})$ as long as the basis set is specified.

4.3 Representation of operators

An operator in quantum mechanics is a 'thing' that operates on one state to produce another, new, state,

$$|\chi\rangle = A|\psi\rangle \qquad (4.16)$$

where in this case A is our operator. By analogy with our previous discussions on group theory, we might want to think of A as one of the symmetry operations. There are other important examples of operators that we will come to as well.

A *linear* operator is one that obeys Eq. 4.16 in addition to

$$A(a_1|\psi_1\rangle + a_2|\psi_2\rangle) = a_1 A|\psi_1\rangle + a_2 A|\psi_2\rangle \qquad (4.17)$$

The product of two linear operators A and B is defined as

$$(AB)|\psi\rangle = A(B|\psi\rangle) \qquad (4.18)$$

where B acts first on $|\psi\rangle$ to give the ket $B|\psi\rangle$; A then acts on this ket. This is exactly analogous to our experience with symmetry operations.

We have a particular interest in *Hermitian* operators in quantum mechanics. Briefly, a Hermitian operator is equal to the complex conjugate of its transpose. See Appendix 4 for more details.

The *commutator* of two operators A and B is defined as

$$[A, B] = AB - BA \tag{4.19}$$

When $[A, B]$ is zero, the two operators A and B are said to commute. When we discuss the Schrödinger equation in Ch. 5, we will see that both the Hermitian nature of the Hamiltonian operator, as well as the operators with which it commutes, have important physical consequences.

Remember that what we actually measure is not a wavefunction or a ket, but a probability, or the ket preceded by a bra,

$$\langle \rho | \chi \rangle = \langle \rho | A | \psi \rangle \tag{4.20}$$

The right-hand side of Eq. 4.20 is called the *matrix element* of the operator A. Note that the matrix element is a number.

Using the fact that we can expand each state vector in an orthonormal basis set, we can write

$$\langle \rho | A | \psi \rangle = \sum_{i,j} \langle \rho | \phi_i \rangle \langle \phi_i | A | \phi_j \rangle \langle \phi_j | \psi \rangle \tag{4.21}$$

The matrix elements are often written as $A_{ij} \equiv \langle \phi_i | A | \phi_j \rangle$.

The quantities A_{ij} are literally matrix elements in the sense that, since they have two indices, they are the components of a square matrix,

$$\begin{pmatrix} A_{11} & A_{12} & \cdots & A_{1j} & \cdots \\ A_{21} & A_{22} & \cdots & A_{2j} & \cdots \\ \vdots & \vdots & \vdots & \vdots & \vdots \\ A_{i1} & A_{i2} & \vdots & A_{ij} & \cdots \\ \vdots & \vdots & \vdots & \vdots & \vdots \end{pmatrix}$$

We usually take the first index i to represent the row index while the second index j represents the column index. Any ket basis vector $|\phi_j\rangle$ can be written as a column vector,

$$\begin{pmatrix} \phi_1 \\ \phi_2 \\ \vdots \\ \phi_j \\ \vdots \end{pmatrix}$$

and, hence, any bra basis vector $\langle \phi_i |$ can be written as a row vector,

$$(\phi_1^* \, \phi_2^* \, \cdots \, \phi_i^* \cdots)$$

The scalar product $\langle \rho | \chi \rangle$ is thus the result of multiplying a row vector by a column vector,

$$\langle \rho | \chi \rangle \;=\; (c_1^* \phi_1^* \, c_2^* \phi_2^* \, \cdots) \begin{pmatrix} c_1 \phi_1 \\ c_2 \phi_2 \\ \vdots \end{pmatrix} \tag{4.22}$$

$$=\; c_1^* b_1 |\phi_1|^2 + c_2^* b_2 |\phi_2|^2 + \ldots \tag{4.23}$$

which is indeed a constant. The jth column of the matrix for the operator A is made up of the components in the $\{|\phi_i\rangle\}$ basis of the product $A|\phi_j\rangle$ of the basis vector $|\phi_j\rangle$.

Exercise 4.6   Using the closure relation, Eq. 4.15, find the matrix that represents the product of the two operators A and B in the basis $\{|\phi_i\rangle\}$.

Matrix elements will be extremely important in what follows on electronic structure, since the matrix elements of the Hamiltonian operator $\mathcal{H}$ are the energy levels associated with our system.

Problems

4.1   If ψ is an unnormalized wavefunction, and N is a constant such that $N\psi$ is normalized, express $|N|$ in terms of ψ.

4.2   What must be the value of the constant α in order that the function $\psi = \iota\alpha xy^2 z^3$ is normalized in the interval $0 \leq x \leq a$, $0 \leq y \leq b$, and $0 \leq z \leq c$?

4.3   Which of the following are sets of linearly independent functions? (a) x, x^2, x^6; (b) $8, x, x^2, 3x^2 - 1$; (c) $\sin x$, $\cos x$; (d) $\sin z$, $\cos z$, $\tan z$; (e) $\sin x$, $\cos x$, $e^{\iota x}$; (f) $\sin^2 x$, $\cos^2 x$, 1; (g) $\sin^2 x$, $\sin^2 y$, 1.

4.4   Exercise 4.5.

4.5   Consider a three-dimensional Hilbert space with three orthonormal basis states $|e_1\rangle, |e_2\rangle$, and $|e_3\rangle$. Let $|a_1\rangle, |a_2\rangle$, and $|a_3\rangle$ be another set of three orthonormal basis states. Write down the transformation that relates the first set of basis states to the second set of basis states in terms of bras and kets.

4.6   Show that $(A + B)^2 = (B + A)^2$ for any two operators A and B, which can be linear or nonlinear. Under what condition is $(A + B)^2 = A^2 + 2AB + B^2$?

4.7 Consider a physical system whose three-dimensional space is spanned by the orthonormal basis formed by the three kets $|u_1\rangle, |u_2\rangle$, and $|u_3\rangle$. In the basis of these three vectors, the operators A and B are:

$$A = \hbar\omega_0 \begin{pmatrix} 1 & 0 & 0 \\ 0 & -1 & 0 \\ 0 & 0 & -1 \end{pmatrix}$$

$$B = b \begin{pmatrix} 1 & 0 & 0 \\ 0 & 0 & 1 \\ 0 & 1 & 0 \end{pmatrix}$$

where ω_0 and b are constants.

(a) Are A and B Hermitian?

(b) Do A and B commute? Use two different techniques to answer this question.

(c) Given the following two wavefunctions expressed in the basis $|u_1\rangle, |u_2\rangle, |u_3\rangle$,

$$\begin{aligned} |\psi_1\rangle &= i|u_1\rangle + c_1|u_2\rangle + c_2|u_3\rangle \\ |\psi_2\rangle &= c_3|u_1\rangle - 2i|u_2\rangle \end{aligned}$$

find the constants c_1, c_2, and c_3 so that these functions are normalized and orthogonal.

4.8 The translation operator is defined by

$$\mathscr{T}_d f(x) \equiv f(x+d) \tag{P4.1}$$

where $f(x)$ is an unknown function. Is $\mathscr{T}_d$ linear? Evaluate $(\mathscr{T}_d^2 - 3\mathscr{T}_d + 2)x^2$.

4.9 In terms of the definition of the commutator of two operators A and B, $[A, B]$, find the following expressions:

(a) $[B, A]$

(b) $[A, (B+C)]$

(c) $[A, BC]$

(d) $[A, [B, C]]$

5

The Schrödinger equation

5.1 Introduction

The mechanics of macroscopic particles are governed by Newton's three laws of motion. However, for microscopic particles like electrons, physicists in the first quarter of the 20th century showed that classical mechanics broke down. Quantum mechanics was introduced to describe the behavior of systems associated with a small de Broglie wavelength, $\lambda = h/p$, where h is Planck's constant and p is the particle momentum. We can see that the de Broglie wavelength of macroscopic objects is essentially zero, so that the laws of quantum mechanics in the limit $\lambda \to 0$ reduce to Newton's equations.

As we mentioned in Ch. 4, the real problem was to find or to derive the proper equation that governed the evolution of a quantum system. In 1926, the Austrian physicist Erwin Schrödinger *postulated* that a wavefunction at the present time evolved to one in the future according to

$$\mathscr{H}\Psi = \imath\hbar\frac{\partial \Psi}{\partial t} \tag{5.1}$$

In this equation Ψ is the position- and time-dependent wavefunction introduced in Ch. 4, $\imath = \sqrt{-1}$, $\hbar$ is Planck's constant divided by 2π, and $\mathscr{H}$ is the Hamiltonian operator,

$$\mathscr{H} = -\frac{\hbar^2}{2m}\nabla^2 + V(\vec{r}, t) \tag{5.2}$$

which is the sum of the kinetic and potential energies of the system. We emphasize the word 'postulated' in connection with the Schrödinger equation, since there is no derivation or proof of this equation. Analogies with geometrical optics or classical mechanics can, however, make the equation plausible, but we will not discuss such analogies here.

Suppose I run a batting cage business in which baseballs are hurled at batters at varying speeds from a pitching machine. I want to make certain that the baseballs fired from the pitching machine will generally cross within the batter's strike zone. I wouldn't stay in business very long if there was a large likelihood that my customers would constantly be dodging baseballs whizzing at their heads at 50 mph! Classical mechanics guarantees that I will be able to predict the trajectory of each baseball. However, in quantum mechanics recall that it is the wavefunction $\Psi(\vec{r}, t)$ that contains all possible information about the trajectory of my system. Unlike classical mechanics, quantum mechanics is basically *statistical* in nature.

$$|\Psi(\vec{r}, t)|^2 d\vec{r} \tag{5.3}$$

gives the probability of finding a particle at time t lying within the region of space $\vec{r}$ to $\vec{r} + d\vec{r}$.

Equation 5.1 is rather formidable and, fortunately for our description of the electronic structure of materials, we will deal with the simpler, *time-independent* Schrödinger equation. We now proceed to derive the time-independent equation from the time-dependent one. We will restrict ourselves to the case in which the potential energy is a function only of position, $V(\vec{r})$. With this constraint, substituting Eq. 5.2 into Eq. 5.1 yields

$$i\hbar \frac{\partial \Psi(\vec{r}, t)}{\partial t} = -\frac{\hbar^2}{2m} \nabla^2 \Psi(\vec{r}, t) + V(\vec{r})\Psi(\vec{r}, t) \tag{5.4}$$

We will look for *separable* solutions of the form

$$\Psi(\vec{r}, t) = \psi(\vec{r})f(t) \tag{5.5}$$

Eq. 5.4 thus becomes

$$i\hbar \psi(\vec{r}) \frac{df}{dt} = -\frac{\hbar^2}{2m} f(t) \nabla^2 \psi(\vec{r}) + V(\vec{r})f(t)\psi(\vec{r})$$

or, dividing through by $\psi(\vec{r})f(t)$,

$$i\hbar \frac{1}{f(t)} \frac{df}{dt} = -\frac{\hbar^2}{2m} \frac{1}{\psi(\vec{r})} \nabla^2 \psi(\vec{r}) + V(\vec{r}) \tag{5.6}$$

Notice now, however, that all of the time dependence is on the left-hand side of Eq. 5.6, while the right-hand side contains all the spatial dependence. The only possible solution is for both sides to be equal to a constant that we will suggestively write as $\mathcal{E}$. The left-hand side of Eq. 5.6 thus becomes

$$\frac{df(t)}{f(t)} = -\frac{i}{\hbar} \mathcal{E} dt \tag{5.7}$$

which has solutions of the form

$$\ln f(t) = -\frac{\imath \mathscr{E} t}{\hbar} + C \tag{5.8}$$

where C is an arbitrary constant of integration. We can more generally write Eq. 5.8 as

$$f(t) = e^{-\imath \mathscr{E} t / \hbar} \tag{5.9}$$

where the constant C has been absorbed into $f(t)$ itself.

Returning to the right-hand side of Eq. 5.6, we have

$$-\frac{\hbar^2}{2m} \nabla^2 \psi(\vec{r}) + V(\vec{r}) \psi(\vec{r}) = \mathscr{E} \psi(\vec{r})$$

or, factoring out $\psi(\vec{r})$,

$$\left[-\frac{\hbar^2}{2m} \nabla^2 + V(\vec{r}) \right] \psi(\vec{r}) = \mathscr{E} \psi(\vec{r}) \tag{5.10}$$

Equation 5.10 is the time-independent Schrödinger equation. More importantly, however, it has the mathematical form of an 'eigenvalue' equation in that application of the Hamiltonian operator, $\mathscr{H}$, to the wavefunction $\psi(\vec{r})$ yields a constant, $\mathscr{E}$, times the same wavefunction. $\psi(\vec{r})$ is called an 'eigenfunction' of $\mathscr{H}$ with $\mathscr{E}$ as its 'eigenvalue'.

The eigenvalue $\mathscr{E}$ is extremely important because, as can be deduced from dimensional analysis of Eq. 5.10, it has units of energy. Thus, when the Hamiltonian operates on a wavefunction $\psi(\vec{r})$, it tells us about the energy of the system. Two other things are important about $\mathscr{E}$. Firstly, the time-independent Schrödinger equation possesses solutions only for a restricted set of eigenvalues $\{\mathscr{E}\}$. This set $\{\mathscr{E}\}$ is called the spectrum of $\mathscr{H}$. Secondly, if there is only *one* eigenfunction $\psi(\vec{r})$ that corresponds to an eigenvalue $\mathscr{E}$, then that function $\psi(\vec{r})$ is said to be nondegenerate. On the other hand, if there is more than one eigenfunction that corresponds to any given eigenvalue, then that eigenvalue is said to be degenerate. The *degree* of degeneracy is the number of linearly independent eigenfunctions that correspond to that eigenvalue.

Returning to Eqs. 5.9 and 5.5, solutions to the Schrödinger equation have the form

$$\Psi(\vec{r}, t) = e^{-\imath \mathscr{E} t / \hbar} \psi(\vec{r}) \tag{5.11}$$

Notice that the probability is a real number, since, from Eq. 5.3,

$$\begin{aligned} |\Psi(\vec{r}, t)|^2 \mathrm{d}\vec{r} &= \Psi^*(\vec{r}, t) \Psi(\vec{r}, t) \mathrm{d}\vec{r} \\ &= e^{\imath \mathscr{E} t / \hbar} \psi^*(\vec{r}) e^{-\imath \mathscr{E} t / \hbar} \psi(\vec{r}) \mathrm{d}\vec{r} \end{aligned}$$

$$= \psi^*(\vec{r}\,)\psi(\vec{r}\,)\mathrm{d}\vec{r}$$
$$= |\psi(\vec{r}\,)|^2\mathrm{d}\vec{r} \tag{5.12}$$

Hence, for states of the form of Eq. 5.11, the probability density is independent of time. Such states are called *stationary* states. The study of the electronic structure of materials is devoted to finding *both* the eigenfunctions, $\psi(\vec{r}\,)$, and the eigenvalues, $\mathscr{E}$, for a system for a given potential energy. Solution of an equation when both of these quantities are unknown is clearly a daunting task! This task can be simplified somewhat by making use of the symmetry of the system, to which we will turn presently.

5.2 The Hamiltonian under symmetry operations

We will now discuss how the Hamiltonian transforms under the symmetry operations of a point group. We will see that $\mathscr{H}$ must transform as Γ_1, or the totally symmetric irreducible representation.

Consider the Hamiltonian of an atom with q electrons. We omit spin-dependent terms for clarity.

$$\mathscr{H} = -\frac{\hbar^2}{2m}\sum_{i=1}^{q}\nabla_i^2 - \sum_{i=1}^{q}\frac{Ze^2}{r_i} + \sum_{i\leq j}^{q}\sum_{i=1}^{q}\frac{e^2}{r_{ij}} \tag{5.13}$$

where the first term is the kinetic energy of q electrons of mass m, the second term is the Coulomb interaction of an electron with a nucleus of charge Ze, and the last term is the Coulomb repulsion between all the q electrons.

How does $\mathscr{H}$ change under a rotation through an angle α about the principal axis? Let us check the individual terms in the Hamiltonian. Consider first some arbitrary function $f(\vec{r}\,)$. Denote the original coordinate system by unprimed coordinates and the new, rotated one by primed coordinates. Then,

$$\frac{\partial f}{\partial x'} = \frac{\partial f}{\partial x}\frac{\partial x}{\partial x'} + \frac{\partial f}{\partial y}\frac{\partial y}{\partial y'} \tag{5.14}$$

We need not consider the z contribution, since we are rotating about this axis.

$$\frac{\partial^2 f}{\partial x'^2} = \frac{\partial^2 f}{\partial x^2}\left(\frac{\partial x}{\partial x'}\right)^2 + \frac{\partial^2 f}{\partial y^2}\left(\frac{\partial y}{\partial y'}\right)^2 \tag{5.15}$$

with a similar equation for $\partial^2 f/\partial y'^2$. Therefore,

$$\nabla'^2 f = \left[\left(\frac{\partial x}{\partial x'}\right)^2 + \left(\frac{\partial x}{\partial y'}\right)^2\right]\frac{\partial^2 f}{\partial x^2} + \left[\left(\frac{\partial y}{\partial x'}\right)^2 + \left(\frac{\partial y}{\partial y'}\right)^2\right]\frac{\partial^2 f}{\partial y^2} \tag{5.16}$$

Exercise 5.1   Using the relationship between two coordinate systems re-
                lated by a rotation, show that each term in square brackets
                is equal to 1.

Therefore,

$$\nabla'^2 f = \nabla^2 f \tag{5.17}$$

and the kinetic energy term is unaltered by application of a rotation operation.

The second term in Eq. 5.13 depends only on the distance between an
electron and the origin, r, and this is not changed by a rotation, such that
$r' = r$. The electron–nucleus interaction is thus not affected by a rotation
operation. Finally, the electron–electron interaction term depends only on
the distance between electrons in the system, r_{ij}.

$$
\begin{aligned}
r_{ij}^2 &= (x_i - x_j)^2 + (y_i - y_j)^2 + (z_i - z_j)^2 \\
&= (x_i' - x_j')^2 + (y_i' - y_j')^2 + (z_i' - z_j')^2 \\
&= r_{ij}'^2
\end{aligned}
\tag{5.18}
$$

The final term is thus unchanged by a rotation. Notice that, for a hydrogen
atom, the final term is absent, since we only have one electron.

We can go ahead and calculate the transformation of the Hamiltonian
under other point group operations and we will find that $\mathcal{H}$ is invariant
under all such operations. The set of transformations that leaves $\mathcal{H}$ invariant
is known as the *group of the Schrödinger equation*.

Exercise 5.2   Following the same type of arguments as above, show ex-
                plicitly that $\mathcal{H}$ is unaffected by inversion and by a mirror
                reflection across the xy plane.

We have thus shown that the Hamiltonian transforms into itself under
all symmetry operations. This makes sense, since, if we look at a physical
system from one coordinate system and then from a related system, the
physics cannot change! If $\mathcal{H}$ transforms into itself under every symmetry
operation, then the corresponding character is $+1$; $\mathcal{H}$ thus transforms as
Γ_1, the totally symmetric irreducible representation.

5.3 Eigenfunctions as basis functions

We want to discuss two theorems relating group theory to quantum mechanics. Both of these theorems make use of the fact that $\mathcal{H}$ is invariant under
application of the symmetry operations of a point group.

What is meant by the effect of a transformation on an operator? Some examples of operators are ∇^2, as discussed in the previous section, and, of course, $\mathcal{H}$. Consider a generic operator $\mathcal{O}$ and two functions f and g. Before applying a symmetry operation, f and g are related by $\mathcal{O}$,

$$f = \mathcal{O}g \tag{5.19}$$

while, after application of the symmetry operation,

$$f' = Rf \tag{5.20}$$

$$g' = Rg \tag{5.21}$$

where R is any symmetry operation of the group. It still must be true that

$$f' = \mathcal{O}'g' \tag{5.22}$$

Therefore,

$$\begin{aligned} f' &= Rf \\ &= R\mathcal{O}g \\ &= R\mathcal{O}(R^{-1}R)g \\ &= R\mathcal{O}R^{-1}g' \end{aligned} \tag{5.23}$$

so that the transformed operator is

$$\mathcal{O}' = R\mathcal{O}R^{-1} \tag{5.24}$$

Invariant operators are those for which $\mathcal{O}$ and $\mathcal{O}'$ are indistinguishable.

Now, since the Hamiltonian is an invariant operator,

$$\begin{aligned} \mathcal{H}' &= R\mathcal{H}R^{-1} \\ &= \mathcal{H} \end{aligned} \tag{5.25}$$

The set of symmetry operations $\{R\}$ that have this property forms a group. This set $\{R\}$ is also said to *commute* with the Hamiltonian.

Exercise 5.3   Show that the set of symmetry operations $\{R\}$ forms a group.

We thus have a group G of symmetry operations under which $\mathcal{H}$ is invariant. Associated with the Hamiltonian is a set of eigenfunctions $\{\psi_i\}$ and eigenvalues $\{\mathscr{E}_i\}$.

Theorem 5.1 *Eigenfunctions that have the same eigenvalue form a basis of a representation of G.*

Table 5.1. *The character table for the point group* C_{2v}

C_{2v}	E	C_2	σ_v	σ_v'		
A_1	1	1	1	1	z	x^2, y^2, z^2
A_2	1	1	-1	-1	R_z	xy
B_1	1	-1	1	-1	x, R_y	xz
B_2	1	-1	-1	1	y, R_x	yz

A more restrictive version of this theorem states that a nondegenerate eigenfunction forms a basis for irreducible representations of the point group of the molecule or crystal. We will examine this case first for clarity. Consider, then, that there is only one ψ_i for each eigenvalue, $\mathcal{E}_i$. Apply a symmetry operation to both sides of the time-independent Schrödinger equation,

$$R\mathcal{H}\psi_i = \mathcal{E}_i R\psi_i \qquad (5.26)$$

or, since R and $\mathcal{H}$ commute (Eq. 5.25),

$$\mathcal{H}R\psi_i = \mathcal{E}_i R\psi_i \qquad (5.27)$$

Thus, the product $R\psi_i$ itself is an eigenfunction also with eigenvalue $\mathcal{E}_i$. Since ψ_i is normalized, we require $R\psi_i$ to be normalized as well,

$$R\psi_i = \pm 1\psi_i \qquad (5.28)$$

By applying each of the operations in the group to an eigenfunction ψ_i belonging to a nondegenerate eigenvalue, we generate a representation of the group with each matrix $\Gamma_i(R)$ equal to ± 1. Since the representations are one-dimensional, they are clearly also irreducible.

As an example of the value of this theorem, consider the point group C_{2v}. The character table for C_{2v} is reproduced in Table 5.1. As a specific example, take the H_2O molecule. The symmetry operations of the point group are $\{E, C_2, \sigma_v, \sigma_v'\}$. Applying each symmetry operation to the H_2O molecule has the effect shown in Fig. 5.1. From the figure, with the coordinate axes centered on the oxygen atom, we can see that $Ep_x = +1p_x$, $C_2p_x = -1p_x$, $\sigma_v p_x = -1p_x$, and $\sigma_v' p_x = +1p_x$. The results for p_y and p_z follow similarly. Therefore, we say that the eigenfunction p_x of the O atom generates the representation B_1 with characters $\{+1, -1, +1, -1\}$. Examination of the character table verifies that x generates B_1. We conclude immediately that since there are only one-dimensional representations of C_{2v}, every eigenfunction must be nondegenerate.

Let us now examine the case when the eigenfunctions are degenerate. Let

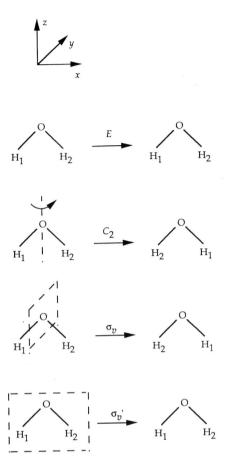

Fig. 5.1. The effects of the symmetry operations of C_{2v} on the H_2O molecule.

ψ_i for $i=1, 2, 3, \ldots, l$ be all the orthonormal eigenfunctions with eigenvalue $\mathscr{E}_i$. Construct a linear combination of these *degenerate* eigenfunctions,

$$\Psi = \sum_{i=1}^{l} a_i \psi_i \qquad (5.29)$$

Let Ψ be normalized,

$$\langle \Psi | \Psi \rangle = 1 \qquad (5.30)$$

or

$$\sum_i a_i^* a_i = 1 \qquad (5.31)$$

Exercise 5.4   Show that this last statement is true provided the degenerate eigenfunctions are orthonormal.

Now, by hypothesis,

$$\mathscr{H}\Psi = \mathscr{E}_i\Psi \tag{5.32}$$

or, for each term in the expansion,

$$\mathscr{H}\psi_i = \mathscr{E}_i\psi_i \tag{5.33}$$

For any symmetry operation R,

$$\begin{aligned} R\mathscr{H}\Psi &= \mathscr{H}(R\Psi) \\ &= \mathscr{E}_i(R\Psi) \end{aligned} \tag{5.34}$$

and

$$\mathscr{H}(R\psi_i) = \mathscr{E}_i(R\psi_i) \tag{5.35}$$

The transformed eigenfunction $R\psi_i$ is also an eigenfunction with the same eigenvalue as ψ_i. In general, $R\psi_i$ must be a linear combination of the l original degenerate eigenfunctions,

$$R\psi_i = \sum_{j=1}^{l} \psi_j(\Gamma(R))_{ij} \tag{5.36}$$

Using all of the symmetry operations of the group $\{A, B, ...\}$, we generate a set of matrices $\Gamma(A)$, $\Gamma(B)$,... and this set forms a representation of G. Theorem 5.2 below will have some other things to say about the nature of this representation.

Consider an ammonia molecule, NH_3, shown in Fig. 5.2. This molecule has the point group C_{3v}, the character table for which is shown in Table 5.2. The symmetry operations of the point group are $\{E, 2C_3, 3\sigma_v\}$. Considering a coordinate system located on the nitrogen atom, we apply each symmetry operation (or one from each class) to the NH_3 molecule. The situation is slightly more complicated now than in the previous example, because we can see that the p_x and p_y atomic orbitals get mixed up with one another: $Ep_x = +1p_x$, $Ep_y = +1p_y$; $C_3p_x = \frac{1}{2}(-p_x - \sqrt{3}p_y)$, $C_3p_y = \frac{1}{2}(\sqrt{3}p_x - p_y)$; $\sigma_v p_x = +1p_x$, $\sigma_v p_y = -1p_y$. Writing p_x and p_y together as a column vector

$$\begin{pmatrix} p_x \\ p_y \end{pmatrix}$$

we can show that the associated characters are $\{2, -1, 0\}$, which is the E representation.

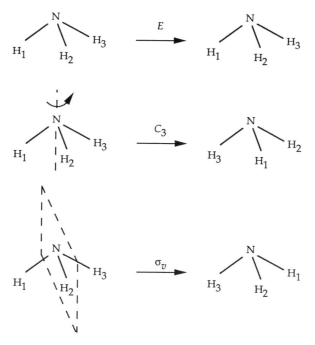

Fig. 5.2. The effects of the symmetry operations of C_{3v} on the NH_3 molecule.

Table 5.2. *The character table for the point group C_{3v}*

C_{3v}	E	$2C_3$	$3\sigma_v$		
A_1	1	1	1	z	$x^2 + y^2, z^2$
A_2	1	1	-1	R_z	
E	2	-1	0	$(x, y), (R_x, R_y)$	$(x^2 - y^2, xy), (xz, yz)$

Theorem 5.2 *Eigenfunctions that are partners of an irreducible representation have the same eigenvalue.*

First of all, what are *partners* of an irreducible representation? Let us consider an irreducible representation Γ_m of dimension l. Then, if the functions $\psi_1^m, \psi_2^m, ..., \psi_l^m$ form a basis for the Γ_m representation, they are said to be *partners* of the Γ_m irreducible representation. Following up on our example of NH_3 will make this clear. Since p_x and p_y *together* generate the E representation, they are partners. Of course, any linear combination of p_x and p_y will also transform as partners of the E representation.

Now, a symmetry operation operating on a partner of a basis of an irreducible representation will result, in general, in a linear combination of all the other orthogonal partners. All the partners of the irreducible representation must be included,

$$R\psi_i = \sum_{j=1}^{l} \psi_j (\Gamma(R))_{ji} \qquad (5.37)$$

and each is an eigenfunction with the same eigenvalue, or energy. Thus, p_x and p_y are degenerate.

One of the consequences of Theorem 5.2 can be seen immediately. Imagine that an atom sits inside a crystal at a site with C_{3v} point symmetry. We need not specifically identify the atom, but suppose that it has an electron configuration, so that its valence electrons occupy p-type orbitals. We know from Fig. 5.2 that x and y transform as partners and, hence, from Theorem 5.2 *the atomic orbitals p_x and p_y must have the same eigenvalue.* The energy levels associated with these wavefunctions are degenerate. Further examination of the C_{3v} character table shows us that the $d_{x^2-y^2}$ and d_{xy} orbitals are degenerate, as are d_{xz} and d_{yz}. If other degeneracies do exist, they occur with a very small probability and are defined as *accidental.*

Exercise 5.5   By examining the character tables in Appendix 3 you can see that the C_{2v} point group is a subgroup of the C_{4v} point group. In turn, C_{4v} is a subgroup of D_{4h}, which is a subgroup of O_h. Using the appropriate character tables, find out how the d wavefunctions transform under each of these groups and identify the degeneracy.

Group theory is a powerful tool for determining the degeneracy of energy levels, but it does not tell us which level is higher in energy than another. For example, we know that the three-fold degeneracy of the p orbitals is broken in a C_{3v} symmetric site. But is the p_z orbital lower in energy than the p_x and p_y orbitals, or *vice versa*? To answer this question, we need to solve the Schrödinger equation. Still, we have learned a great deal about electronic structure using only symmetry.

5.4 More on matrix elements

We will now present a very powerful statement regarding the nature of matrix elements of the Hamiltonian. Specifically, we will be able to identify *by inspection* when Hamiltonian matrix elements are zero or nonzero. This is extremely important, since the Hamiltonian matrix elements are just the energies of our system.

Consider the time-independent Schrödinger equation,

$$\mathcal{H}\psi_i = \mathcal{E}\psi_i \tag{5.38}$$

Multiply on the left side of this equation by ψ_j and integrate both sides over all space to obtain

$$\int \psi_j \mathcal{H}\psi_i d\vec{r} = \mathcal{E} \tag{5.39}$$

where we have already made the implicit assumption that the wavefunctions are orthonormal. This equation is an explicit expression for the energy of interaction between a state $|\psi_i\rangle$ and a state $|\psi_j\rangle$. It would be helpful if we knew without a great deal of computational effort when this eigenvalue was to be zero. Fortunately, this can be done by examining the irreducible representations to which the two wavefunctions belong.

First note that the Hamiltonian operator must have the full symmetry of the molecule or crystal. That this must be true can be seen by considering the fact that the matrix elements of $\mathcal{H}$ are simply real numbers which *must* be independent of application of any symmetry operation. We have shown this above formally when we showed that the Hamiltonian generated the totally symmetric irreducible representation. The symmetry of the integrand then depends on the representations contained in what is called the *direct product* of the representations for ψ_i and ψ_j. Without going into any details, it can be shown that the totally symmetric representation can occur in the direct product representation only if ψ_i and ψ_j both belong to the *same* irreducible representation. We may then state the following theorem:

Theorem 5.3 *The expectation value of the Hamiltonian between two states $|\psi_i\rangle$ and $|\psi_j\rangle$ may be nonzero only if the two states belong to the same irreducible representation of the point group of the molecule or crystal.*

This theorem is often invoked when trying to identify nonzero spectral transition probabilities and we will use it to determine nonzero matrix elements of the Hamiltonian for band structure calculations.

Problems

5.1   Consider two time-independent wavefunctions ψ_1 and ψ_2, each of which is an eigenfunction of $\mathcal{H}$. That is,

$$\mathcal{H}\psi_1 = \mathcal{E}_1\psi_1$$
$$\mathcal{H}\psi_2 = \mathcal{E}_2\psi_2$$

(a) Show that the linear combination of $\psi_1 + \psi_2$ is *not* an eigenfunction of $\mathcal{H}$ if $\mathcal{E}_1 \neq \mathcal{E}_2$.

(b) Under what special condition is $\psi_1 + \psi_2$ an eigenfunction of $\mathcal{H}$? Are all linear combinations of ψ_1 and ψ_2 eigenfunctions under the same condition?

5.2   Solve the time-independent Schrödinger equation for a particle in free space, *i.e.*, it feels no potential energy. By solving, we mean write down the appropriate Schrödinger equation, find the wavefunctions and eigenvalues. Note that you will find *two* solutions. By considering the behavior for large r, you can reject one of your solutions. How will you normalize this wavefunction? What does the probability density tell you about this wavefunction?

5.3 $\mathcal{H}$ is a Hermitian operator such that $\mathcal{H}^\dagger = \mathcal{H}$.

(a) Show that the eigenvalues must be real.

(b) Show that two eigenfunctions of $\mathcal{H}$ corresponding to different eigenvalues are orthogonal.

5.4   Exercise 5.2.

5.5   Exercise 5.4.

5.6   Exercise 5.5.

5.7   Consider a particle in a one-dimensional system with Hamiltonian

$$\mathcal{H} = \frac{p^2}{2m} + V(x) \qquad\qquad (P5.1)$$

where $V(x) = \lambda x^n$.

(a) Determine the commutator $[\mathcal{H}, xp]$.

(b) Are there any stationary states for this potential? If so, calculate the mean value of the kinetic energy $<T>$ and the mean value of the potential energy $<V>$ in these states. What is the relationship between these two quantities?

This result is called the *virial theorem*.

5.8   Consider a linear triatomic molecule consisting of atoms A, B, and C as shown in Fig. 5.3 containing *one* electron. Wavefunctions that are localized on each one of these atoms are denoted $|\phi_A\rangle$, $|\phi_B\rangle$, and $|\phi_C\rangle$, respectively. Assume that these three basis functions are orthonormal.

If we do not allow the electron to hop between neighboring atoms, then the energy of this system is described by the Hamiltonian $\mathcal{H}_0$, which has eigenfunctions $|\phi_A\rangle$, $|\phi_B\rangle$, and $|\phi_C\rangle$ with eigenvalue $\mathcal{E}_0$. If the electron is allowed to hop between neighboring atoms, then the coupling is described by an additional contribution to the Hamiltonian $\mathcal{H}_1$, defined by:

Fig. 5.3.   Linear triatomic molecule used for Problem 5.8.

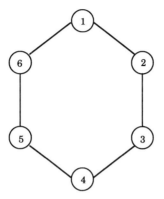

Fig. 5.4.   Planar hexagonal molecule consisting of six identical atoms A used for Problem 5.9.

$$\mathcal{H}_1|\phi_A\rangle = -a|\phi_B\rangle$$
$$\mathcal{H}_1|\phi_B\rangle = -a|\phi_A\rangle - a|\phi_C\rangle$$
$$\mathcal{H}_1|\phi_C\rangle = -a|\phi_B\rangle$$

$$(P5.2)$$

where a is a real positive constant.

   (a) Calculate the eigenvalues and wavefunctions of the full Hamiltonian $\mathcal{H} = \mathcal{H}_0 + \mathcal{H}_1$.
   (b) The electron is localized on atom A at $t = 0$. Describe qualitatively the localization of the electron at subsequent times. Are there any values of time t for which the electron is perfectly localized at atom A, B, or C?

5.9   Imagine a molecule that consists of six identical atoms A which sit at the corners of a perfect planar hexagon as depicted in Fig. 5.4. The atoms are labeled with subscripts 1 through 6 only to distinguish them. Consider placing one electron into this system and that it can be localized on any one of the atoms. The state $|\phi_n\rangle$ is the state which is localized on the nth atom, $n = 1, 2, \ldots, 6$. The collection of states $|\phi_n\rangle$ are orthonormal.

(a) Define the operator R by the following relations:

$$R|\phi_1\rangle = |\phi_2\rangle$$
$$R|\phi_2\rangle = |\phi_3\rangle$$
$$\vdots$$
$$R|\phi_6\rangle = |\phi_1\rangle$$

(P5.3)

Find the eigenvalues and eigenfunctions of the operator R. Show that these eigenfunctions span the space of the states.

(b) If the electron cannot hop from one atom to either of its neighbors, then the system is described by a Hamiltonian $\mathcal{H}_0$ with eigenvalue $\mathcal{E}_0$ for the six states $|\phi_n\rangle$. By adding a perturbation to $\mathcal{H}_0$, we can allow for electron hopping. The total Hamiltonian is thus $\mathcal{H} = \mathcal{H}_0 + \mathcal{H}_1$, where $\mathcal{H}_1$ is defined by

$$\mathcal{H}_1|\phi_1\rangle = -a|\phi_6\rangle - a|\phi_2\rangle$$
$$\mathcal{H}_2|\phi_1\rangle = -a|\phi_1\rangle - a|\phi_3\rangle$$
$$\vdots$$
$$\mathcal{H}_6|\phi_1\rangle = -a|\phi_5\rangle - a|\phi_1\rangle$$

Show that R commutes with the total Hamiltonian $\mathcal{H}$. From this result find the eigenfunctions and eigenvalues of $\mathcal{H}$. Is the electron localized in these states? Can you apply these considerations to the benzene molecule?

6

The hydrogen atom

6.1 Introduction

Shown in part (a) of Fig. 6.1 is a contour plot of the calculated charge density of a Be atom and an O atom separated by 5 Å. Recall that the charge density is just $|\psi(\vec{r})|^2$ for stationary states of the time-independent Schrödinger equation. The contours in this figure are lines where the charge density is equal. Contours closest to the center of each atom have the highest probability for finding the electrons in the system. Note particularly the circular (or spherical, since this is just a two-dimensional slice through a three-dimensional object) symmetry about the center of each atom. At 5 Å separation, the atoms are far enough apart such that they do not interact with one another.

At 3.5 Å separation between the Be and the O atom, part (b) of Fig. 6.1, the charge density contours begin to overlap *both* atoms. This is the first indication that a bond is being formed between the two atoms. At the equilibrium interatomic separation of 1.3 Å, part (c) of Fig. 6.1 shows that there is a great deal of charge density shared between the two atoms, but that the higher-density contours are skewed toward the O atom. The skewness is a reflection of the greater electronegativity of the O atom compared with Be. What is most striking when comparing all three of these figures, however, is that there are *remnants* of the *atomic* charge densities in the *molecular* charge density.

Figure 6.2 shows on the left an ideal plane in a crystal of graphite. The atoms are at the vertices of regular hexagons. The calculated charge density contour plot is shown in the accompanying figure on the right. One can identify regions of high charge density (where the contour spacing is small) with the location of the atoms. The charge density is small in regions not occupied by atoms, such as at the center of the sixfold planar rings.

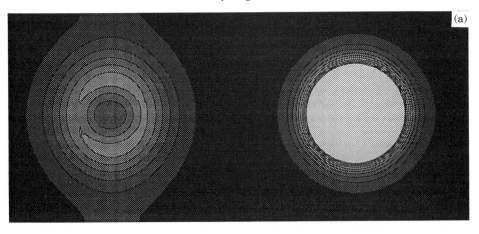

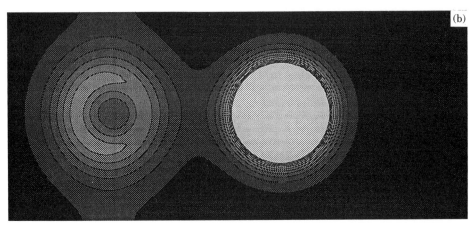

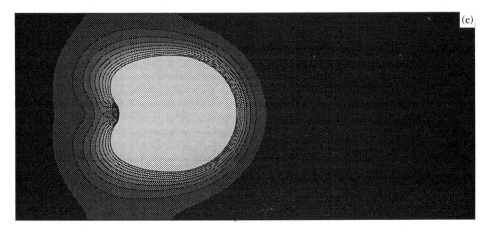

Fig. 6.1. Charge density contour plot for BeO at several internuclear separations. (a) 5.0 Å, minimum contour density is 0.000 Å^{-3}, maximum contour density is 6.044 Å^{-3}; (b) 3.5 Å, minimum contour density is 0.000 Å^{-3}, maximum contour density is 6.025 Å^{-3}; (c) 1.3 Å, minimum contour density is 0.000 Å^{-3}, maximum contour density is 6.535 Å^{-3}. Figures courtesy of P.J. Samsel.

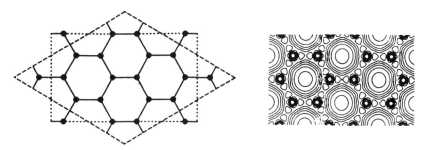

Fig. 6.2. Single plane of graphite (left) and accompanying calculated charge density. From E.F. Kaxiras and K.C. Pandey, *Phys. Rev. Lett.* **61**, 2693 (1988). Used with permission.

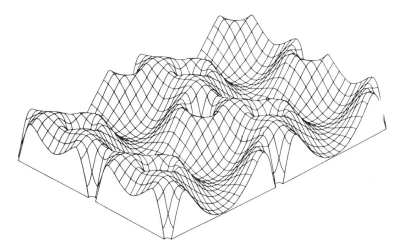

Fig. 6.3. Relative electron density on a (110) plane in silicon. Relative electron density is measured vertically from the basal plane. From W. A. Harrison, *Pseudopotentials in the Theory of Metals* (W.A. Benjamin, Amsterdam, 1966), p. 224. Used with permission.

Figure 6.3 shows the calculated charge density surface plot for the (110) plane in Si. The atomic nuclei are located at the deep minima. Note the high charge density between atoms, in the bonds, and the lack of charge elsewhere.

On the other hand, if we consider a charge density surface plot for a metal, we might find something like that shown in Fig. 6.4 for a crystal of face-centered-cubic Al. The surface is plotted for a (110) plane. The contrast between Fig. 6.4 and Figs. 6.1, 6.2, and 6.3 is striking: no longer is

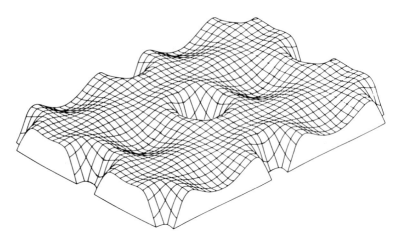

Fig. 6.4.   Relative electron density on a (110) plane in fcc aluminum. Relative electron density is measured vertically from the basal plane. From W. A. Harrison, *Pseudopotentials in the Theory of Metals* (W.A. Benjamin, Amsterdam, 1966), p. 225. Used with permission.

the charge density lumped between nearest-neighbor atoms. Now it is much more uniformly distributed throughout an entire atomic plane.

The importance of Figs. 6.1, 6.2, and 6.3 lies in our ability to recognize the *atomic* charge density even within the context of a molecule or a crystal. This identification lends strong intuitive support to the tight-binding model for electronic structure that we introduced in Ch. 4.† The wavefunction for an entire system is written as a linear combination of basis functions,

$$|\psi\rangle = \sum_i \langle \phi_i | \psi \rangle | \phi_i \rangle \tag{6.1}$$

The basis functions in the case of the tight-binding approximation are just the atomic orbitals of the constituent atoms. Chemists call this method 'Linear Combination of Atomic Orbitals' (LCAO), for obvious reasons. However, because the charge density is more uniformly distributed in a metal, we cannot identify individual atomic orbitals. We will see in a later chapter that a more appropriate basis set for the wavefunction of a metal consists of plane waves.

Because the tight-binding approximation can be applied successfully to many different molecules and crystals, it is important that we understand

† The tight-binding method has received more rigorous theoretical justification in recent years. See W.M.C. Foulkes and R. Haydock, *Phys. Rev. B* **39**, 12520 (1989).

the origin and nature of the individual atomic orbitals. These atomic orbitals are just solutions of the hydrogen atom, to which we now turn.

6.2 The reduced mass problem

To find the wavefunctions of the hydrogen atom, we must solve the time-independent Schrödinger equation for a single electron moving in the potential of a single proton. This potential energy belongs to a class of potentials of the central-force type wherein the interaction depends only on the interparticle separation.

Let us postulate that we can extend the one-particle wavefunction introduced in Ch. 5 to a two-particle wavefunction that depends on the coordinates of both the electron, $\vec{r}_e$, and the nucleus, $\vec{r}_n$. The time-independent Schrödinger equation is thus given by

$$\left[-\frac{\hbar^2}{2m_e} \nabla_e^2 - \frac{\hbar^2}{2m_n} \nabla_n^2 + V(\vec{r}_e, \vec{r}_n) \right] \psi(\vec{r}_e, \vec{r}_n) = \mathscr{E} \psi(\vec{r}_e, \vec{r}_n) \qquad (6.2)$$

where m_e (m_n) is the mass of the electron (nucleus) and ∇_e^2 (∇_n^2) is the Laplacian for the electron (nuclear) coordinates.

Now, we have already stipulated that the electron–nucleus interaction potential depends only on their separation, $V(\vec{r}_e, \vec{r}_n) = V(|\vec{r}_e - \vec{r}_n|)$. This allows us to make the tremendous simplification of introducing *relative* coordinates and *center of mass* coordinates,

$$\begin{aligned} \vec{r} &= \vec{r}_e - \vec{r}_n \\ M\vec{R} &= m_e \vec{r}_e + m_n \vec{r}_n \end{aligned} \qquad (6.3)$$

where $M = m_e + m_n$. We can rewrite Eq. 6.2 in terms of the new coordinates,

$$\left[-\frac{\hbar^2}{2\mu} \nabla_r^2 - \frac{\hbar^2}{2M} \nabla_R^2 + V(|\vec{r}|) \right] \psi(\vec{r}, \vec{R}) = \mathscr{E} \psi(\vec{r}, \vec{R}) \qquad (6.4)$$

where

$$\mu = \frac{m_e \cdot m_n}{m_e + m_n} \qquad (6.5)$$

is the reduced mass of the electron–nucleus system. Since $m_n \approx 1837 m_e$, $\mu \approx m_e$. Relative coordinates describe the motion of the electron *relative* to the nucleus, while center of mass coordinates describe the motion of the center of mass of the entire electron–nucleus system.

We can separate the wavefunction $\psi(\vec{r}, \vec{R})$ into the product of a function that depends only on the relative coordinates, $\psi_r(\vec{r})$, and a function that depends only on the center of mass coordinates, $\psi_{cm}(\vec{R})$. This is entirely analogous to what we did in Ch. 5 when separating the wavefunction $\Psi(\vec{r}, t)$ into the product $f(t)\psi(\vec{r})$.

Inserting this separable wavefunction into Eq. 6.4 yields two equations,

$$-\frac{\hbar^2}{2M}\nabla_R^2\psi_{cm}(\vec{R}) = \mathscr{E}_{cm}\psi_{cm}(\vec{R})$$

$$-\frac{\hbar^2}{2\mu}\nabla_r^2\psi_r(\vec{r}) + V(r)\psi_r(\vec{r}) = \mathscr{E}\psi_r(\vec{r}) \qquad (6.6)$$

Exercise 6.1   Show this last step.

The first equation is simply the Schrödinger equation for a free particle of mass M, since no potential energy term appears. $\mathscr{E}_{cm}$ is just the kinetic energy associated with translation of the center of mass through free space. The solutions to the second equation are for the relative electron–nucleus motion. In the next section we will turn our attention to finding the eigenvalues and eigenfunctions for the electron moving relative to a stationary nucleus.

Exercise 6.2   Find solutions for the center of mass equation. What does the form of these solutions tell you about the entire system?

6.3 Separation of variables in the equation of relative motion

In Ch. 5 we examined the transformation of the Hamiltonian under symmetry operations. There we introduced the potential energy for an electron interacting with a nucleus of charge Ze,

$$V(r) = -\frac{Ze^2}{r} \qquad (6.7)$$

The other term in the potential energy was the electron–electron repulsion, which is absent for the H atom. Because the potential energy is spherically symmetric, it is most convenient to recast the Schrödinger equation in spherical coordinates (r, θ, ϕ) as defined in Fig. 6.5. In this case, Eq. 6.6 takes the form

$$-\frac{\hbar^2}{2\mu r^2}\left[\frac{\partial}{\partial r}\left(r^2\frac{\partial\psi}{\partial r}\right) + \frac{1}{\sin\theta}\frac{\partial}{\partial\theta}\left(\sin\theta\frac{\partial\psi}{\partial\theta}\right) + \frac{1}{\sin^2\theta}\frac{\partial^2\psi}{\partial\phi^2}\right] - \frac{Ze^2}{r}\psi = \mathscr{E}\psi$$

$$(6.8)$$

where we have dropped all subscripts on the wavefunction and $\psi = \psi(r, \theta, \phi)$.

We will try separable functions as solutions to Eq. 6.8,

$$\psi(r, \theta, \phi) = R(r)\Theta(\theta)\Phi(\phi) \qquad (6.9)$$

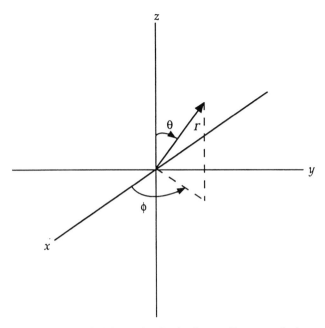

Fig. 6.5. Definition of spherical coordinates r, θ, ϕ.

Inserting this wavefunction into Eq. 6.8 and following some simple re-arrangement, we find

$$\frac{\sin^2 \theta}{R} \frac{\mathrm{d}}{\mathrm{d}r} \left(r^2 \frac{\mathrm{d}R}{\mathrm{d}r} \right) + \frac{\sin \theta}{\Theta} \frac{\mathrm{d}}{\mathrm{d}\theta} \left(\sin \theta \frac{\mathrm{d}\Theta}{\mathrm{d}\theta} \right) + \frac{1}{\Phi} \frac{\mathrm{d}^2 \Phi}{\mathrm{d}\phi^2}$$
$$+ \frac{2\mu r^2 \sin^2 \theta}{\hbar^2} \left(\mathscr{E} + \frac{Ze^2}{r} \right) = 0 \qquad (6.10)$$

Exercise 6.3   Show this last step.

The third term in Eq. 6.10 involves only ϕ, while the other terms are independent of variations in ϕ. Therefore, let

$$\frac{\mathrm{d}^2 \Phi}{\mathrm{d}\phi^2} = A\Phi \qquad (6.11)$$

where A is a constant. Inserting Eq. 6.11 back into Eq. 6.10 yields an equation that is separable in the r and θ coordinates. Following some straightforward manipulations, we arrive at:

$$\frac{1}{\sin \theta} \frac{\mathrm{d}}{\mathrm{d}\theta} \left(\sin \theta \frac{\mathrm{d}\Theta}{\mathrm{d}\theta} \right) + \frac{A}{\sin^2 \theta} \Theta = -B\Theta \qquad (6.12)$$

Table 6.1. *Selected normalized* $\Phi_m(\phi)$ *functions*

m	$\Phi_m(\phi)$
0	$1/\sqrt{2\pi}$
+1	$(1/\sqrt{2\pi})e^{i\phi}$
−1	$(1/\sqrt{2\pi})e^{-i\phi}$
+2	$(1/\sqrt{2\pi})e^{2i\phi}$
−2	$(1/\sqrt{2\pi})e^{-2i\phi}$

$$\frac{\hbar^2}{2\mu r^2}\frac{d}{dr}\left(r^2\frac{dR}{dr}\right) + \left(\frac{Ze^2}{r}\right)R - \left(\frac{B\hbar^2}{2\mu r^2}\right) = -\mathscr{E}R \qquad (6.13)$$

Exercise 6.4   Find Eqs. 6.12 and 6.13 from Eqs. 6.10 and 6.11.

Equations 6.11, 6.12, and 6.13 constitute the three separated equations of the Schrödinger equation for an electron in the spherically symmetric potential of a nucleus. Explicit solution of these three equations is accomplished by solving Eq. 6.11 first, using A in Eq. 6.12 to solve it, and then finally solving the radial equation, Eq. 6.13. This process is well detailed in any quantum mechanics textbook, so that we will not present it here.

6.4 Complete solutions for the hydrogen atom

The ϕ-dependent equation has a solution of the form

$$\Phi_m(\phi) = \frac{1}{\sqrt{2\pi}}e^{im\phi} \qquad (6.14)$$

where $m = 0, \pm1, \pm2, ...$, and the factor $1/\sqrt{2\pi}$ results from normalization. The eigenvalues A are quantized, since $A = -m^2$. The integer m is known as the magnetic quantum number. A selection of ϕ-dependent solutions is given in Table 6.1.

Solutions to the θ-dependent equation have the form

$$\Theta_{lm}(\theta) = \left[\frac{(2l+1)}{2}\frac{(l-|m|)!}{(l+|m|)!}\right]^{\frac{1}{2}} P_l^{|m|}(\cos\theta) \qquad (6.15)$$

where $l=0, 1, 2, ..., \infty$, m is restricted to positive and negative integers such that $m=0, \pm1, \pm2, ..., \pm l$, and $P_l^{|m|}(\cos\theta)$ are the associated Legendre polynomials. Table 6.2 shows selected normalized solutions to the θ-dependent equation. l is known as the azimuthal quantum number.

Table 6.2. Selected normalized $\Theta_{lm}(\theta)$ functions

l	m	$\Theta_{lm}(\theta)$
0	0	$\sqrt{2}/2$
1	0	$(\sqrt{6}/2)\cos\theta$
1	± 1	$(\sqrt{3}/2)\sin\theta$
2	0	$(\sqrt{10}/4)(3\cos^2\theta - 1)$
2	± 1	$(\sqrt{15}/2)\sin\theta\cos\theta$
2	± 2	$(\sqrt{15}/4)\sin^2\theta$

Table 6.3. Selected normalized $R_{nl}(r)$ functions

n	l	$R_{nl}(r)$
1	0	$2\left(\dfrac{Z}{a_0}\right)^{\frac{3}{2}} e^{-Zr/a_0}$
2	0	$\left(\dfrac{Z}{2a_0}\right)^{\frac{3}{2}}\left(2 - \dfrac{Zr}{a_0}\right) e^{-Zr/2a_0}$
2	1	$\dfrac{1}{\sqrt{3}}\left(\dfrac{Z}{2a_0}\right)^{\frac{3}{2}}\left(\dfrac{Zr}{a_0}\right) e^{-Zr/2a_0}$
3	0	$\dfrac{2}{3}\left(\dfrac{Z}{3a_0}\right)^{\frac{3}{2}}\left(3 - \dfrac{2Zr}{a_0} + \dfrac{2Z^2r^2}{9a_0^2}\right) e^{-Zr/3a_0}$
3	1	$\dfrac{2\sqrt{2}}{9}\left(\dfrac{Z}{3a_0}\right)^{\frac{3}{2}}\left(\dfrac{2Zr}{a_0} - \dfrac{Z^2r^2}{3a_0^2}\right) e^{-Zr/3a_0}$
3	2	$\dfrac{4}{27\sqrt{10}}\left(\dfrac{Z}{3a_0}\right)^{\frac{3}{2}}\left(\dfrac{Z^2r^2}{a_0^2}\right) e^{-Zr/3a_0}$

The solutions to the radial equation are quite complicated,

$$R_{nl}(r) = -\left\{\left(\frac{2Z}{na_0}\right)^3 \frac{(n-l-1)!}{2n[(n+l)!]^3}\right\}^{\frac{1}{2}} \rho^l e^{-\frac{\rho}{2}} L_{n+1}^{2l+1}(\rho) \qquad (6.16)$$

where

$$\rho = \left(\frac{2Z}{na_0}\right) r \qquad (6.17)$$

a_0 is the Bohr radius, Z is the atomic number, l is the azimuthal quantum number from Eq. 6.15, n is the principal quantum number, and $L_{n+1}^{2l+1}(\rho)$ are the associated Laguerre polynomials. Since l is allowed the values 0, 1, 2, ..., n is restricted to positive integers, $n=1, 2, 3, ..., \infty$. The restriction on n means that l is restricted to values 0, 1, 2, ..., $(n-1)$. Some selected normalized radial eigenfunctions are shown in Table 6.3.

The total wavefunction for the hydrogen atom is expressed as

$$\psi_{nlm} = R_{nl}(r) \cdot \Theta_{lm}(\theta) \cdot \Phi_m(\phi) \tag{6.18}$$

and the associated eigenvalue is

$$\mathcal{E} = -\frac{2\pi^2 \mu Z^2 e^4}{n^2 h^2} \tag{6.19}$$

According to this last equation, the electron energy depends only on the principal quantum number, n, but not on the azimuthal or magnetic quantum numbers, l and m. Since n is associated with the radial equation, we infer that the electron's energy depends only on its separation from the nucleus and is angularly independent. This is in accord with our original supposition that the potential is spherically symmetric.

Shown in Figs. 6.6–6.14 are three-dimensional plots of the hydrogen atomic orbitals that we will use most frequently. Depicted there are contour surfaces of constant probability density. Such surfaces are found by evaluating the probability integral for a preselected value. For example,

$$\int |\psi|^2 d\vec{r} = 0.95 \tag{6.20}$$

means that there is a 95% probability of locating the electron within this surface. The nomenclature is simple. The principal quantum number is always given explicitly and, since $\mathcal{E} \propto -(1/n^2)$, the larger the value of n, the higher the energy. The azimuthal quantum number is given a letter: $l=0$ corresponds to s, $l=1$ corresponds to p, $l=2$ corresponds to d, *etc.* Lastly, for reasons of convenience, the magnetic quantum number is often replaced by a linear combination of functions with the same l value. Thus we have p_x, p_y, and p_z for the $l=1$ orbitals and others as listed in Table 6.4. The '+' and '−' signs on different lobes of orbitals other than s orbitals indicate the *parity* of the wavefunction. The parity of an orbital is what happens to it under inversion. The parity of an orbital is given by $(-1)^l$, so s orbitals are even, p orbitals are odd, d orbitals are even, and so on. These trends are reflected in Figs. 6.6–6.14. The radial dependence for the s-type wavefunctions is shown in Fig. 6.15, for the p-type wavefunctions in Fig. 6.16, and for the d-type wavefunctions in Fig. 6.17.

Problems

6.1 The eigenfunctions of the H atom have an associated parity: they are either even or odd, depending on the value of $(-1)^l$, where l is the

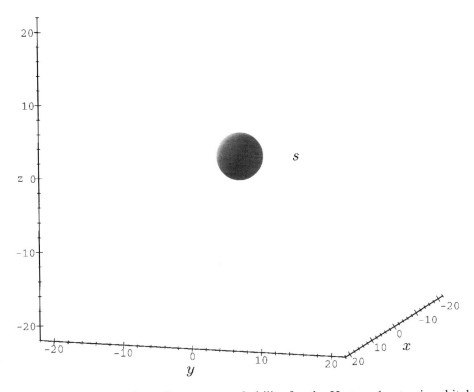

Fig. 6.6. Contour surface of constant probability for the H atom 1*s* atomic orbital.

Table 6.4. *Angular wavefunctions for atomic orbitals*

Letter type	Full polynomial	Simplified polynomial	Normalizing factor	Angular function
s			$\sqrt{\pi}/2$	
	z		$\sqrt{\frac{3}{\pi}}/2$	$\cos\theta$
p	x		$\sqrt{\frac{3}{\pi}}/2$	$\sin\theta\cos\phi$
	y		$\sqrt{\frac{3}{\pi}}/2$	$\sin\theta\sin\phi$
	$2z^2 - x^2 - y^2$	z^2	$\sqrt{\frac{5}{\pi}}/4$	$(3\cos^2\theta - 1)$
	xz		$\sqrt{\frac{15}{\pi}}/2$	$\sin\theta\cos\theta\cos\phi$
d	yz		$\sqrt{\frac{15}{\pi}}/2$	$\sin\theta\cos\theta\sin\phi$
	$x^2 - y^2$		$\sqrt{\frac{15}{\pi}}/4$	$\sin^2\theta\cos 2\phi$
	xy		$\sqrt{\frac{15}{\pi}}/4$	$\sin^2\theta\sin 2\phi$

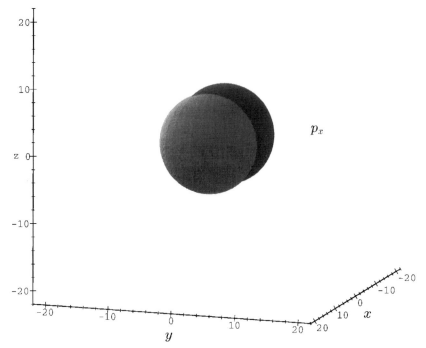

Fig. 6.7. Contour surface of constant probability for the H atom $2p_x$ atomic orbital.

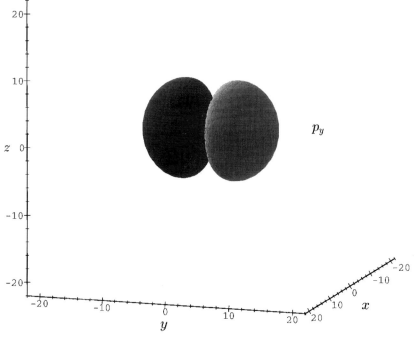

Fig. 6.8. Contour surface of constant probability for the H atom $2p_y$ atomic orbital.

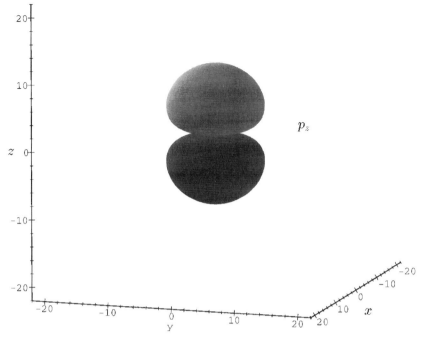

Fig. 6.9. Contour surface of constant probability for the H atom $2p_z$ atomic orbital.

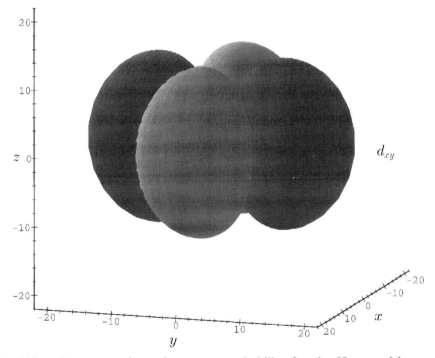

Fig. 6.10. Contour surface of constant probability for the H atom $3d_{xy}$ atomic orbital.

The hydrogen atom

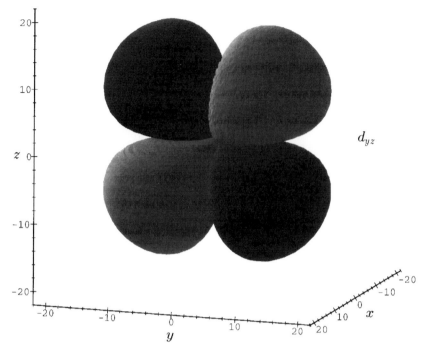

Fig. 6.11.   Contour surface of constant probability for the H atom $3d_{yz}$ atomic orbital.

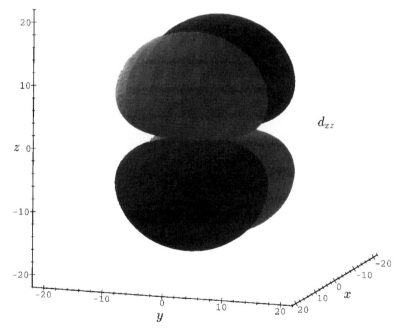

Fig. 6.12.   Contour surface of constant probability for the H atom $3d_{xz}$ atomic orbital.

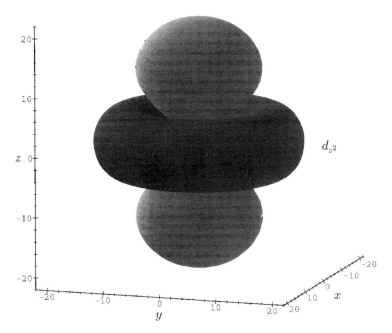

Fig. 6.13.   Contour surface of constant probability for the H atom $3d_{z^2}$ atomic orbital.

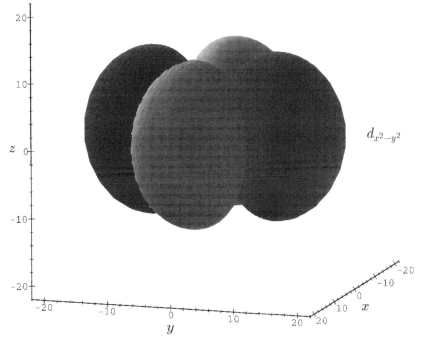

Fig. 6.14.   Contour surface of constant probability for the H atom $3d_{x^2-y^2}$ atomic orbital.

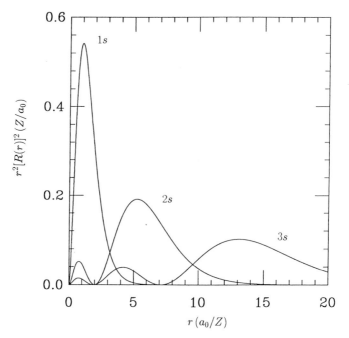

Fig. 6.15. Radial dependence for *s*-type atomic orbitals.

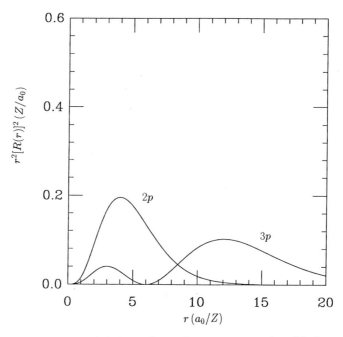

Fig. 6.16. Radial dependence for *p*-type atomic orbitals.

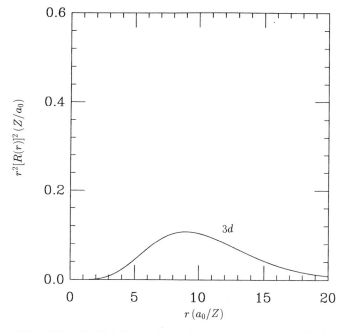

Fig. 6.17. Radial dependence for *d*-type atomic orbitals.

orbital angular momentum quantum number. Parity will turn out to be an extremely important concept when we begin discussing bonding in complex systems.

In quantum mechanics, there is actually a parity operator, Π, that has no classical analog.

(a) Using the definition of Π,

$$\Pi|\vec{r}\rangle = |-\vec{r}\rangle \tag{P6.1}$$

or

$$\Pi f(\vec{r}) = f(-\vec{r}) \tag{P6.2}$$

find the eigenvalues and eigenfunctions of the parity operator.

(b) Recall that we found that a set of symmetry operations $\{R\}$ commuted with the Hamiltonian, *i.e.*,

$$R\mathscr{H} = \mathscr{H}R \tag{P6.3}$$

One important theorem from quantum mechanics states that if two operators commute, they have a common set of eigenfunctions. Under what conditions do the Hamiltonian,

$$\mathcal{H} = -\frac{\hbar}{2m}\nabla^2 + V(r) \tag{P6.4}$$

and Π commute? What does this say about the hydrogen atom wavefunctions?

6.2 Where is the probability density a maximum for the H atom ground state?

6.3 A stationary state wavefunction, ψ, is an eigenfunction of the Hamiltonian operator, $\mathcal{H} = \mathcal{T} + \mathcal{V}$, where $\mathcal{T}$ is the kinetic energy contribution and $\mathcal{V}$ is the potential energy contribution. It is sometimes erroneously believed that ψ is also an eigenfunction of $\mathcal{T}$ *and* of $\mathcal{V}$. For the ground state of the H atom, verify that ψ is not an eigenfunction of either $\mathcal{T}$ or $\mathcal{V}$, but of $(\mathcal{T} + \mathcal{V})$. Compute $<\mathcal{T}>$, $<\mathcal{V}>$, and $<\mathcal{T}>/<\mathcal{V}>$.

6.4 Using Eq. 6.19, calculate the energy difference between $n=1$ and $n=2$ levels, $n=2$ and $n=3$ levels, and $n=3$ and $n=4$ levels. To what wavelengths do these transitions correspond? What can you say about the trend in energy level separation as $n \to \infty$?

7

Diatomic molecules

7.1 Homonuclear diatomic molecules

In Ch. 6 we went into considerable detail to find the wavefunctions and energies for the hydrogen atom. This is one of the few exactly soluble problems in quantum mechanics. From spectroscopic experiments we have gained confirmation that this is indeed the correct electronic structure of the H atom. However, we clearly want to explore more complex systems. What are we up against? As already indicated in Ch. 5, for a many-electron atom we will have to contend with electron–electron repulsion. Also, for molecules, we will have to deal with *atom–atom* interactions. Studying the electronic structure of crystals involves both of these additional problems as well as *defects* inherent in crystals, such as vacancies, interstitials, and surfaces. The electronic structure of a real material is quite complicated.

However, we already have some insight into how we might proceed. In Ch. 4 we introduced the tight-binding approximation where a many-atom (or many-electron) wavefunction is written as a linear combination of single-atom wavefunctions,

$$|\psi\rangle = \sum_i \langle\phi_i|\psi\rangle|\phi_i\rangle \tag{7.1}$$

where we will use the solutions to the H atom from Ch. 6 as our basis states, $\{|\phi_i\rangle\}$. We turn now to solving for the electronic structure of the simplest *molecule* that we can imagine, the homonuclear diatomic. This system will give us intuition about how we will eventually treat solids, whether crystalline or not, as *very large molecules*.

For specificity, let us concentrate on the H_2 molecule, although what we will present can be generalized to any homonuclear diatomic. When we studied the electronic structure of the H atom, we found it extremely helpful to separate coordinates into relative coordinates and center of mass coordinates. We then concentrated solely on solving for the energies and wavefunctions of the electron relative to the nucleus. We will make use of a similar approximation in studying molecules. We assume that the motion of the very light electrons is much more rapid than that of the relatively heavy and sluggish nuclei, so that the electrons are able to adjust immediately to any change in nuclear positions. This is known as the *Born–Oppenheimer approximation*. We therefore focus our attention on solving for the energies and wavefunctions of the electrons moving in the potential of essentially static nuclei.

From the periodic table we know that the electronic structure of each constituent H atom is $1s^1$, where the notation indicates that the principal quantum number is 1, the azimuthal (or orbital angular momentum) quantum number is $l = 0\,(s)$, and that this state is populated by one electron (the superscript). Therefore, let the molecular wavefunction be written as a combination of the two atomic wavefunctions,

$$|\psi\rangle = c_1|\phi_{1s}(1)\rangle + c_2|\phi_{1s}(2)\rangle \tag{7.2}$$

where c_1 and c_2 are constants to be determined and $|\phi_{1s}(1)\rangle$ and $|\phi_{1s}(2)\rangle$ are the atomic wavefunctions centered on atoms 1 and 2, respectively.

Inserting this expression into the Schrödinger equation, we obtain

$$\begin{aligned}
\mathcal{H}|\psi\rangle &= \mathcal{H}(c_1|\phi_{1s}(1)\rangle + c_2|\phi_{1s}(2)\rangle) \\
&= \mathcal{E}(c_1|\phi_{1s}(1)\rangle + c_2|\phi_{1s}(2)\rangle)
\end{aligned} \tag{7.3}$$

where the eigenvalue is taken to be $\mathcal{E}$. Take the overlap of the Schrödinger equation sequentially with the bras $\langle\phi_{1s}(1)|$ and $\langle\phi_{1s}(2)|$ to yield two equations,

$$\begin{aligned}
c_1\langle\phi_{1s}(1)|\mathcal{H}|\phi_{1s}(1)\rangle + c_2\langle\phi_{1s}(1)|\mathcal{H}|\phi_{1s}(2)\rangle = \\
\mathcal{E}(c_1\langle\phi_{1s}(1)|\phi_{1s}(1)\rangle + c_2\langle\phi_{1s}(1)|\phi_{1s}(2)\rangle) \\
c_1\langle\phi_{1s}(2)|\mathcal{H}|\phi_{1s}(1)\rangle + c_2\langle\phi_{1s}(2)|\mathcal{H}|\phi_{1s}(2)\rangle = \\
\mathcal{E}(c_1\langle\phi_{1s}(2)|\phi_{1s}(1)\rangle + c_2\langle\phi_{1s}(2)|\phi_{1s}(2)\rangle)
\end{aligned} \tag{7.4}$$

We can simplify these equations by recognizing that the basis functions are normalized,

$$\langle\phi_{1s}(1)|\phi_{1s}(1)\rangle = \langle\phi_{1s}(2)|\phi_{1s}(2)\rangle = 1 \tag{7.5}$$

but not necessarily orthogonal,

$$\langle\phi_{1s}(1)|\phi_{1s}(2)\rangle = \langle\phi_{1s}(2)|\phi_{1s}(1)\rangle \equiv S_{12} \neq 0 \tag{7.6}$$

The term S_{12} is known as the overlap integral and is a measure of the nonorthogonality of the orbitals on atomic centers 1 and 2. In addition, let us introduce a similar shorthand notation to that at the end of Ch. 4,

$$\langle \phi_{1s}(1)|\mathcal{H}|\phi_{1s}(2)\rangle = \mathcal{H}_{12} \tag{7.7}$$

and likewise for the other matrix elements.

We can re-express Eqs. 7.4 as

$$
\begin{aligned}
c_1\mathcal{H}_{11} + c_2\mathcal{H}_{12} &= \mathcal{E}(c_1 + c_2 S_{12}) \\
c_1\mathcal{H}_{21} + c_2\mathcal{H}_{22} &= \mathcal{E}(c_1 + c_2 S_{21})
\end{aligned}
\tag{7.8}
$$

These two equations are called the *secular equations* for the H_2 molecule, and we know from linear algebra that a nontrivial solution for c_1 and c_2 exists only if the determinant of the coefficients is zero,

$$
\begin{vmatrix}
\mathcal{H}_{11} - \mathcal{E} & \mathcal{H}_{12} - \mathcal{E}S_{12} \\
\mathcal{H}_{21} - \mathcal{E}S_{21} & \mathcal{H}_{22} - \mathcal{E}
\end{vmatrix} = 0
$$

Expanding the determinant,

$$(\mathcal{H}_{11} - \mathcal{E})(\mathcal{H}_{22} - \mathcal{E}) - (\mathcal{H}_{21} - \mathcal{E}S_{21})(\mathcal{H}_{12} - \mathcal{E}S_{12}) = 0 \tag{7.9}$$

We can simplify Eq. 7.9 further by recognizing that $\mathcal{H}_{11} = \mathcal{H}_{22} \equiv \mathcal{E}_0$, the energy eigenvalue of a single H atom. In addition, it can be shown that $\mathcal{H}_{12} = \mathcal{H}_{21}^*$ and that $S_{12} = S_{21}$.

Exercise 7.1   Making use of the fact that $\mathcal{H}$ is a *Hermitian* operator, show that $\mathcal{H}_{12} = \mathcal{H}_{21}^*$ and $S_{12} = S_{21}$.

But $\mathcal{H}_{12}$ is real, as can be seen from the fact that the atomic orbitals are real. Such *off-diagonal* matrix elements are often denoted as β. We introduce a further simplification by making the *neglect of overlap* approximation, in which all overlap integrals are set equal to zero, $S_{12}=0$. This sounds very drastic, but it turns out that the neglect of overlap approximation works surprisingly well and does not seem to introduce errors of any greater magnitude than those that are already present. Therefore, the solutions to the secular Eqs. 7.8 are of the form

$$\mathcal{E} = \mathcal{E}_0 \pm \beta \tag{7.10}$$

Clearly one state is lower in energy than the other, but the ordering depends on the sign of β.

To answer this last question, consider the Hamiltonian for a *single* electron in the H_2 molecule,†

$$\mathcal{H} = -\frac{\hbar^2}{2m}\nabla^2 + V_1(r) + V_2(r) \tag{7.11}$$

where $V_1(r)$ ($V_2(r)$) is the Coulomb interaction between the electron and nucleus 1 (2). Now,

$$\begin{aligned}
\mathcal{H}_{12} &= \langle\phi_{1s}(1)|\mathcal{H}|\phi_{1s}(2)\rangle \\
&= \langle\phi_{1s}(1)| -\frac{\hbar^2}{2m}\nabla^2 + V_1(r) + V_2(r)|\phi_{1s}(2)\rangle \\
&= \langle\phi_{1s}(1)| \left[-\frac{\hbar^2}{2m}\nabla^2 + V_1(r)\right]|\phi_{1s}(2)\rangle \\
&\quad + \langle\phi_{1s}(1)|V_2(r)|\phi_{1s}(2)\rangle
\end{aligned} \tag{7.12}$$

The term in square brackets is zero because that part of the Hamiltonian operating on the bra $\langle\phi_{1s}(1)|$ just yields the appropriate eigenvalue multiplied by $\langle\phi_{1s}(1)|$ and the overlap S_{12} integral has been assumed to equal zero. The remaining term sandwiches an *attractive* and, hence, negative, potential between the two basis states. Since the basis states are $1s$ states and are everywhere positive, $\mathcal{H}_{12} = \beta < 0$.

Let us suggestively relabel our energy eigenvalues from Eq. 7.10 as

$$\begin{aligned}
\mathcal{E}_b &= \mathcal{E}_0 + \beta \\
\mathcal{E}_a &= \mathcal{E}_0 - \beta
\end{aligned} \tag{7.13}$$

where $\mathcal{E}_b < \mathcal{E}_a$. The corresponding normalized eigenfunctions are

$$\begin{aligned}
|\psi_b\rangle &= \frac{1}{\sqrt{2}}(|\phi_{1s}(1)\rangle + |\phi_{1s}(2)\rangle) \\
|\psi_a\rangle &= \frac{1}{\sqrt{2}}(|\phi_{1s}(1)\rangle - |\phi_{1s}(2)\rangle)
\end{aligned} \tag{7.14}$$

Exercise 7.2   Find the normalized wavefunctions of Eq. 7.14 by substituting the energies, Eq. 7.13, back into Eqs. 7.4 and determining the constants c_1 and c_2.

Sketches of these *molecular* wavefunctions, often called molecular orbitals by analogy with atoms, are shown in Fig. 7.1. The reason for the subscripts 'b' and 'a' now becomes apparent. In the top figure, the electron wavefunction, and, hence, the charge density, is piled up between the two nuclei. But this represents formation of a bond! Chemists often call a bond with this

---

† I am indebted to A.P. Sutton for this argument. See A.P. Sutton, *Electronic Structure of Materials* (Oxford University Press, Oxford, 1993), p. 27.

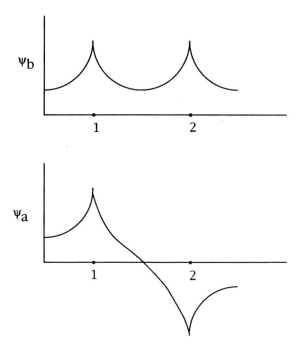

Fig. 7.1. The bonding (top) and antibonding (bottom) wavefunctions for a homonuclear diatomic molecule.

geometry a 'σ' bond. In the bottom figure, there is actually a region of zero probability density between the nuclei: electron density is clustered more around the atoms than between them. This is called an 'antibonding' state.

In Fig. 7.2 we show a schematic of the energy eigenvalues of the H_2 molecule. The bonding state is lower in energy than the isolated atom by an amount β. The reason for this energy lowering is because in the bonding state each electron feels the attraction of *two* nuclei. At the same time, the antibonding state is higher in energy than the isolated atom by β. This may seem a bit strange at first: why isn't the energy of the antibonding state just $\mathscr{E}_0$? It has been pointed out by several authors that as a result of the node between the two H atoms, the electrons in the antibonding state will have a higher kinetic energy because they are confined to the region immediately surrounding their respective atoms. However, this argument violates the virial theorem, which states that the kinetic energy be exactly half the magnitude of the potential energy. (See Problems, Ch. 5.) The antibonding state has higher energy than the separated nuclei because it cannot take

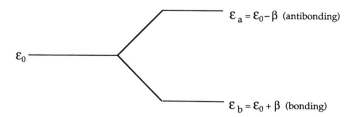

Fig. 7.2. Schematic diagram showing the splitting between the bonding and antibonding states in a homonuclear diatomic molecule. The two states are derived from a pair of atomic energy levels with energy $\mathscr{E}_0$.

advantage of the attraction that lowers the energy of the bonding state. The splitting between the two states is thus 2β.

What can group theory tell us about this simple molecule? The point group of H_2 is $D_{\infty h}$ if we take the z axis as coincident with the bond axis. The notation $D_{\infty h}$ means that rotations of C_ϕ through all possible angles ϕ about the z axis are symmetry operations (also denoted as $C_\infty(\phi)$) and that twofold rotations C_2' exist about any axis bisecting the plane perpendicular to the z axis. Because the two atoms in the diatomic molecule are identical, there is also a horizontal mirror plane halfway between the two atoms. Combining the reflection operation σ_h with a rotation by π about the z axis gives the inversion operation, i. Combining inversion with the C_2' operation gives a reflection operation denoted as σ_v. Finally, this group, which is one of the *linear groups*, also contains improper rotations which are obtained by combining i with $C_\infty(\phi)$. The character table for this group is shown in Table 7.1. The prefix ∞ before the C_2', reflection, and improper rotation operations indicates that they are infinite in number.

The irreducible representations for this point group are labeled in two ways. The first label is the familiar Mulliken notation. The second label is always an upper case Greek letter by analogy with the atomic scheme for orbitals (s, p, d, ...). For the case of H_2 we have only two electrons, which, we know from the Pauli exclusion principle, can occupy the lowest-energy orbital with their spins opposed. Arguments concerning the sum of the electron spins that are beyond the scope of our discussion here show that we must have a Σ state. From our treatment above and Fig. 7.1, we know that the bonding state is even under inversion and reflection and, hence, must be the Σ_g^+ state. Likewise, the antibonding state must be the Σ_u^+ state. From the theorems in Ch. 5 linking group theory and quantum mechanics,

Table 7.1. The character table for the linear group $D_{\infty h}$

	E	$2C_\infty(\phi)$		$\infty\sigma_v$	i	$2S_\infty(\phi)$		$\infty C_2'$			
$A_{1g}(\Sigma_g^+)$	1	1	$\cdots$	1	1	1	$\cdots$	1			x^2+y^2, z^2
$A_{2g}(\Sigma_g^-)$	1	1	$\cdots$	-1	1	1	$\cdots$	-1	R_z		
$E_{1g}(\Pi_g)$	2	$2\cos\phi$	$\cdots$	0	2	$-2\cos\phi$	$\cdots$	0	(R_x, R_y)	(xz, yz)	
$E_{2g}(\Delta_g)$	2	$2\cos2\phi$	$\cdots$	0	2	$2\cos2\phi$	$\cdots$	0		(x^2-y^2, xy)	
$\cdots$	$\vdots$	$\vdots$		$\vdots$	$\vdots$	$\vdots$		$\vdots$			
E_{ng}	2	$2\cos n\phi$	$\cdots$	0	2	$(-1)^n 2\cos2\phi$	$\cdots$	0			
$\cdots$	$\vdots$	$\vdots$		$\vdots$	$\vdots$	$\vdots$		$\vdots$			
$A_{1u}(\Sigma_u^+)$	1	1	$\cdots$	1	-1	-1	$\cdots$	-1	z		
$A_{2u}(\Sigma_u^-)$	1	1	$\cdots$	-1	-1	-1	$\cdots$	1			
$E_{1u}(\Pi_u)$	2	$2\cos\phi$	$\cdots$	0	-2	$2\cos\phi$	$\cdots$	0	(x, y)		
$E_{2u}(\Delta_u)$	2	$2\cos2\phi$	$\cdots$	0	-2	$-2\cos2\phi$	$\cdots$	0			
$\cdots$	$\vdots$	$\vdots$		$\vdots$	$\vdots$	$\vdots$		$\vdots$			
E_{nu}	2	$2\cos n\phi$	$\cdots$	0	-2	$(-1)^{n+1} 2\cos2\phi$	$\cdots$	0			
$\cdots$	$\vdots$	$\vdots$		$\vdots$	$\vdots$	$\vdots$		$\vdots$			

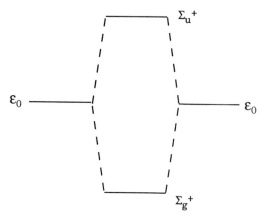

Fig. 7.3. A molecular orbital diagram for the homonuclear diatomic molecule. The bonding (Σ_u^+) and antibonding (Σ_g^-) states are labeled according to the group $D_{\infty h}$.

since both the Σ_g^+ and Σ_u^+ irreducible representations are one-dimensional, the accompanying states must be nondegenerate.

A revised diagram of the energy eigenvalues is shown in Fig. 7.3. Here the noninteracting atomic orbital energies are shown on the extreme left and right of the diagram. Since the two atoms are identical, these energies are identical. The molecular energy levels are shown in the center of the diagram with their appropriate labels attached. Chemists call this a *molecular orbital diagram*. Notice that the splitting between the two states is equal; the bonding level drops in energy as much as the antibonding level rises. By the Pauli exclusion principle, we know that each state can accommodate two electrons. A total of four electrons can occupy the molecular orbitals of H_2. We will return to the consequences of occupation later in this chapter.

What about this curious off-diagonal matrix element β? We will leave the details of the calculation to a problem at the end of the chapter and simply state that the off-diagonal Hamiltonian matrix elements are called 'hopping integrals' because β literally tells us about the transfer of electrons from atom 1 to atom 2. The larger the magnitude of β, the lower the bonding state energy, and the higher the probability that an electron 'tunnels' from atom 1 to atom 2.

The last thing that we will explore for the homonuclear diatomic molecule is what happens if the valence electrons occupy orbitals other than s orbitals. Consider diatomic boron, B_2. The electron configuration of atomic boron is $1s^2 2s^2 2p^1$. If we assume that the $2s$ and $2p$ electrons do not interact, then the valence electron configuration is just $2p^1$. The molecular wavefunction is identical in form with Eq. 7.2, so that the secular determinant is the same and, hence, the eigenvalues are the same as Eq. 7.13. Similarly, the wavefunctions are identical in form with Eq. 7.14.

There is one crucial difference, however. How do we know *which p* orbital the electron in each B atom occupies? It makes quite a difference! To see this, let us sketch the possible scenarios. Take the bond axis as the z axis in Fig. 7.4. Suppose first that the electron on each B atom occupies the p_z orbital. Then the bonding and antibonding molecular orbitals formed look like the schematic diagram in the top of Fig. 7.4. Comparing this figure with Fig. 7.1, we can see that overlap of p_z orbitals with their lobes oriented toward one another produces σ bonds.

Next, put the electron in the p_x orbital. The charge density lobes are perpendicular to the bond axis now! The bonding wavefunction in part (b) of Fig. 7.4 still puts charge density in the region between the atomic center; it is just off the bond axis. This is called a π bond.

Exercise 7.3   What if we choose p_y instead of p_x? Because the charge density directly between atoms is smaller for π bonds than for σ bonds, the latter form stronger bonds. One consequence is that, even though the p orbitals are degenerate in the atom, the degeneracy is lifted in the diatomic molecule. The p_z-derived state is lower in energy than the p_x- (or p_y-) derived state.

Exercise 7.4   Is there another way in which you know that the three p orbitals cannot be degenerate in the diatomic molecule? How?

We can construct the same argument for a (hypothetical) diatomic molecule that has one valence d electron. By examining the character table for the linear group $D_{\infty h}$ (Table 7.1), we can show that the d_{z^2} orbital transforms as Σ_g^+, d_{xz} and d_{yz} together transform as Π_g, and $d_{x^2-y^2}$ and d_{xy} together transform as Δ_g. The d_{z^2} combination forms a σ bond, the d_{xz} and d_{yz} combinations form two π bonds, while the $d_{x^2-y^2}$ and d_{xy} combinations form two δ bonds. The last type of bonds are characterized by two nodal planes that contain the bond axis, but are perpendicular to each other. See part (c) of Fig. 7.4. δ bonds are weaker than either π or σ bonds.

7.2 Heteronuclear diatomic molecules

Following the treatment of the H_2 molecule in the previous section, we are now in a position to study a heteronuclear diatomic molecule. The molecular

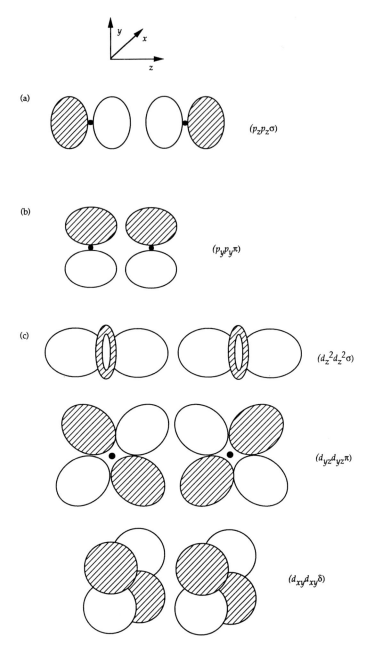

Fig. 7.4. Combination of (a) p_z orbitals to form a σ bond; (b) p_y (or p_x) orbitals to form a π bond; and (c) d_{z^2} to form a σ bond, d_{yz} (or d_{xz}) to form a π bond, and $d_{x^2-y^2}$ (or d_{xy}) to form a δ bond.

wavefunction can once again be written as a sum over atomic wavefunctions on the two constituent atoms, which we will call A and B for the sake of generality,

$$|\psi\rangle = c_1|\phi_A\rangle + c_2|\phi_B\rangle \qquad (7.15)$$

We do not specify the quantum numbers of the basis functions, for generality as well. Letting $\beta = \mathcal{H}_{AB} = \mathcal{H}_{BA} < 0$ and using notation equivalent to that in the previous section, we arrive at the following two equations:

$$\begin{aligned}
c_1\mathcal{H}_{AA} + c_2\beta &= \mathcal{E}(c_1 + c_2 S_{AB}) \\
c_1\beta + c_2\mathcal{H}_{BB} &= \mathcal{E}(c_1 S_{BA} + c_2)
\end{aligned} \qquad (7.16)$$

Exercise 7.5   Derive the secular equations in Eq. 7.16.

The heteronuclear diatomic is slightly more complicated than the homonuclear diatomic because $\mathcal{H}_{AA} \neq \mathcal{H}_{BB}$. Rather, let $\mathcal{H}_{AA} \equiv \mathcal{E}_A$ and $\mathcal{H}_{BB} \equiv \mathcal{E}_B$ and, for the sake of argument, $\mathcal{E}_A > \mathcal{E}_B$. Recall that $\mathcal{E}_A$ and $\mathcal{E}_B$ are the eigenvalues associated with the *noninteracting* atoms A and B. Making the neglect of overlap approximation again, $S_{AB} = 0$, our energy eigenvalues are

$$\begin{aligned}
\mathcal{E}_b &= \alpha - \sqrt{\Delta^2 + \beta^2} \\
\mathcal{E}_a &= \alpha + \sqrt{\Delta^2 + \beta^2}
\end{aligned} \qquad (7.17)$$

where $\alpha = (\mathcal{E}_A + \mathcal{E}_B)/2$ and $\Delta = (\mathcal{E}_A - \mathcal{E}_B)/2$.

Exercise 7.6   Verify this last result.

As a consistency check, if we let $\Delta \to 0$, we recover the eigenvalues of Eq. 7.13. The energy eigenvalues are plotted schematically in Fig. 7.5. Now the splitting between the bonding and antibonding states is $2\sqrt{\Delta^2 + \beta^2}$, which is larger than that for the homonuclear diatomic and increases as Δ increases.

Recall that the wavefunctions for the homonuclear diatomic had equal contributions from each atomic center. This was reflected as a symmetrical charge density, $|c_1|^2 = |c_2|^2 = \frac{1}{2}$. However, because $\mathcal{E}_A > \mathcal{E}_B$ for the heteronuclear diatomic, the charge density is skewed toward atom B, which has the lower-energy atomic basis state. It can be shown that for the *bonding* state,

$$\frac{c_1^2}{c_2^2} = \frac{1}{1 + 2x^2 + 2x\sqrt{1 + x^2}} \qquad (7.18)$$

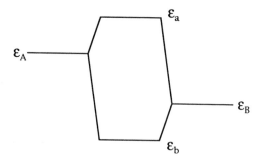

Fig. 7.5. Schematic plot of the molecular energy eigenvalues for the heteronuclear diatomic molecule.

while for the *antibonding* state,

$$\frac{c_1^2}{c_2^2} = \frac{1}{1 + 2x^2 - 2x\sqrt{1 + x^2}} \qquad (7.19)$$

where $x = \Delta/\beta$.

Exercise 7.7   Derive Eqs. 7.18 and 7.19.

These results are plotted schematically in Fig. 7.6. From the figure it can be seen that as $x \to 0$, because either $\Delta \to 0$ or $\beta \to \infty$, the charge density ratio returns to the equal-sharing diatomic case. However, as x increases in magnitude, the charge density skewness increases, with a much higher density on atom B, the lower-energy atom. We recognize this unequal charge-sharing as a polar or ionic bond. But where does the difference $\mathscr{E}_A > \mathscr{E}_B$ originate? It comes from differences in the *electronegativity* of atoms A and B, a concept we introduced in Ch. 1.

The point group for a heteronuclear diatomic differs from the point group of a homonuclear diatomic. The principal axis is still coincident with the bond axis and we still possess the symmetry operation $C_\infty(\phi)$, which is now broken down into classes of rotations by 2ϕ. However, we lose the horizontal mirror plane. This means that, in addition to the identity, the only other symmetry operation is the reflection σ_v. The character table for the $C_{\infty v}$ point group is shown in Table 7.2. The irreducible representations are labeled with upper case Greek letters, as was the case for $D_{\infty h}$. The lowest-energy bonding state generates the Σ^+ irreducible representation and, hence, is nondegenerate.

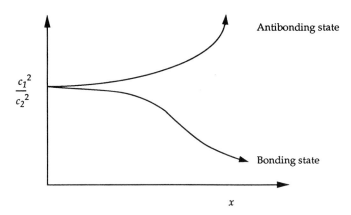

Fig. 7.6. Ratio of the square of the coefficients in front of the two atomic wavefunctions in the heteronuclear diatomic wavefunction as a function of $x = \Delta/\beta$.

Table 7.2. *The character table for the linear group* $C_\infty(\phi)$

	E	$2C_\infty(\phi)$	$2C_\infty(2\phi)$	$\cdots$	$\infty\sigma_v$		
$A_1(\Sigma^+)$	1	1	1	$\cdots$	1	z	$x^2 + y^2, z^2$
$A_2(\Sigma^-)$	1	1	1	$\cdots$	-1	R_z	
$E_1(\Pi)$	2	$2\cos\phi$	$2\cos 2\phi$	$\cdots$	0	$(x, y), (R_x R_y)$	(xz, yz)
$E_2(\Delta)$	2	$2\cos 2\phi$	$2\cos 4\phi$	$\cdots$	0		$(x^2 - y^2, xy)$
$\cdots$	$\cdots$	$\cdots$	$\cdots$	$\cdots$	$\cdots$		
$\cdots$	$\cdots$	$\cdots$	$\cdots$	$\cdots$	$\cdots$		
$E_n(\Delta)$	2	$2\cos n\phi$	$2\cos 2n\phi$	$\cdots$	0		

7.3 Bond energy and bond order

Let us reconsider the formal statement of the eigenvalue problem for a homonuclear diatomic molecule, written now as a matrix equation,

$$\begin{pmatrix} \mathcal{H}_{11} & \mathcal{H}_{12} \\ \mathcal{H}_{21} & \mathcal{H}_{22} \end{pmatrix} \begin{pmatrix} c_1 \\ c_2 \end{pmatrix} = \mathcal{E} \begin{pmatrix} c_1 \\ c_2 \end{pmatrix}$$

Let the eigenvectors be $\begin{pmatrix} c_1^b \\ c_2^b \end{pmatrix}$ for the bonding state and $\begin{pmatrix} c_1^a \\ c_2^a \end{pmatrix}$ for the antibonding state. Multiplying on the left by the row vector $(c_1^{b*}\ c_2^{b*})$ for the bonding equation, we have

$$(c_1^{b*}\ c_2^{b*}) \begin{pmatrix} \mathcal{H}_{11} & \mathcal{H}_{12} \\ \mathcal{H}_{21} & \mathcal{H}_{22} \end{pmatrix} \begin{pmatrix} c_1^b \\ c_2^b \end{pmatrix} = \mathcal{E}_b(c_1^{b*}c_1^b + c_2^{b*}c_2^b)$$

Using the fact that $(c_1^{b*}c_1^b + c_2^{b*}c_2^b) = 1$, we can obtain an expression for $\mathscr{E}_b$ by expanding the left side of the above matrix equation,

$$\mathscr{E}_b = c_1^b c_1^{b*} \mathscr{H}_{11} + c_2^b c_2^{b*} \mathscr{H}_{22} + \{c_1^b c_2^{b*} \mathscr{H}_{12} + c_1^{b*} c_2^b \mathscr{H}_{21}\} \tag{7.20}$$

and likewise for the antibonding state,

$$\mathscr{E}_a = c_1^a c_1^{a*} \mathscr{H}_{11} + c_2^a c_2^{a*} \mathscr{H}_{22} + \{c_1^a c_2^{a*} \mathscr{H}_{12} + c_1^{a*} c_2^a \mathscr{H}_{21}\} \tag{7.21}$$

The first term in Eq. 7.20 is proportional to the probability that an electron spends its time in the vicinity of atom 1. The second term expresses the probability that an electron spends its time in the vicinity of atom 2. Both of these terms are physically reasonable and really have little to do with bonding. The term in braces, the so-called interference term, codifies bonding in the diatomic molecule. This term is proportional to the probability that an electron spends its time *between* the two nuclei. This 'glue' term is called the bond energy.

The bond energy term can be rewritten as $2c_1^b c_2^b \mathscr{H}_{21}$, where we have made use of the fact that the eigenvectors are real and that $\mathscr{H}_{12} = \mathscr{H}_{21}$. The matrix element coefficient, $2c_1^b c_2^b$, is called the *partial bond order*. The *bond order* is the sum of these partial bond orders over all occupied states. It follows that the *bond energy* is just the product of the bond order and the hopping matrix element. Because of the fact that

$$2c_1^b c_2^b + 2c_1^a c_2^a = 0 \tag{7.22}$$

if *all* the bonding and antibonding states are fully occupied, the bond order is zero and there is no net bond in the molecule. This is an important point, to which we will return when we discuss bonding in solids.

Exercise 7.8   Draw a sketch of how the bond order varies in the H_2 molecule as a function of the number of electrons populating the states.

7.4 General secular equation

For the sake of completeness, and as a taste of what is to come, we briefly discuss the more general secular equation. This discussion could also be appropriate for the homo- or heteronuclear diatomic molecules should we want to expand the number of orbitals in the basis set.

Consider that we can write a molecular wavefunction as a sum over a complete set of atomic orbitals,

$$|\psi\rangle = \sum_{i=1}^{\infty} c_i|\phi_i\rangle \qquad (7.23)$$

Substituting this expansion into the Schrödinger equation, multiplying on the left by $\langle\phi_j|$, we obtain

$$\sum_{i=1}^{\infty} c_i[\langle\phi_j|\mathscr{H}|\phi_i\rangle - \mathscr{E}\langle\phi_j|\phi_i\rangle] = 0 \qquad (7.24)$$

for $j = 1, 2, 3, ...$ This infinite set of equations will have a solution for all c_i if the determinant of the coefficients vanishes,

$$\begin{vmatrix} \mathscr{H}_{11} - \mathscr{E}S_{11} & \mathscr{H}_{12} - \mathscr{E}S_{12} & \cdots \\ \mathscr{H}_{21} - \mathscr{E}S_{21} & \mathscr{H}_{22} - \mathscr{E}S_{22} & \cdots \\ \vdots & \vdots & \vdots \end{vmatrix} = 0$$

The solution of this infinite-dimensional secular equation will yield an infinite number of energy eigenvalues for the molecule (or crystal). There will be as many equations as unknown coefficients c_i, so the problem is soluble in principle.

   In practice, this is quite another matter. We are not generally interested in an infinite number of energy eigenvalues, but only the few lower-lying ones. This is because most of the properties of a molecule or a crystal are determined by the highest-lying, or *valence* electrons. So we can truncate the expansion in Eq. 7.23 to only a few dominant terms, neglecting sometimes even lower-lying orbitals if they are far below the valence orbitals. Such orbitals are known as *core* orbitals. Furthermore, there is another approximation, due to Hückel, that, although it appears somewhat arbitrary and extreme, actually works quite well in a number of cases. Namely, we ignore all *off-diagonal* matrix elements that do not connect adjacent atoms. We will make extensive use of this approximation in subsequent chapters.

Problems

7.1   Consider the time-dependent Schrödinger equation for the state $|\psi\rangle$,

$$\mathscr{H}|\psi\rangle = i\hbar\frac{d|\psi\rangle}{dt} \qquad (P7.1)$$

    (a) Obtain two equations for a total wavefunction that is the sum of

Table 7.3. *Bond energy and bond length trends for the first row homonuclear diatomics*

	B_2	C_2	N_2	O_2	F_2	Ne_2
Bond energy (kcal/mol)	69	150	225	118	36	–
Bond length (Å)	1.59	1.24	1.10	1.21	1.44	–

two atomic wavefunctions. Note that your two coefficients will be time-dependent,

$$|\psi\rangle = c_1(t)|\phi_{1s}(1)\rangle + c_2(t)|\phi_{1s}(2)\rangle \tag{P7.2}$$

where the subscript $1s$ refers to the type of atomic orbital and the labels '1' and '2' are for the two atoms in a homonuclear diatomic. Make the same substitutions as was done for the homonuclear diatomic, namely $\mathcal{H}_{11} = \mathcal{H}_{22} = \mathcal{E}_0$, $\mathcal{H}_{12} = \mathcal{H}_{21} = \beta$, and solve for the coefficients $c_1(t)$ and $c_2(t)$.

(b) Now suppose we have the initial condition that $c_1(t = 0) = 1$ and $c_2(t = 0) = 0$, so that the molecule is initially in state $|\phi_{1s}(1)\rangle$. Find the appropriate expression for $c_1(t)$ and $c_2(t)$ using these initial conditions. What is the probability that the molecule is in the state $|\phi_{1s}(2)\rangle$? How does this probability vary with time? What role does β play? From your answers to these questions, it should be clear why β is called the 'hopping integral'.

7.2 (a) Using the concept of bond order and the electronic structure of the first row homonuclear diatomics, explain the trends in Table 7.3.

(b) Magnetic properties of homonuclear diatomics can also be deduced from these simple electronic structure diagrams and a knowledge of Hund's rules. The most important of Hund's rules says that a system will try to maximize its spin; that is, try to keep as many spins as possible parallel. Which of the above homonuclear diatomic molecules would you expect to be magnetic (*i.e.*, have unpaired spins)?

7.3 We have discussed the homonuclear diatomic molecule H_2 within the 'neglect of overlap approximation'. Now consider including the overlap, $S_{12} = \langle\phi_{1s}(1)|\phi_{1s}(2)\rangle \neq 0$. Write down the appropriate molecular wavefunction, set up the secular determinant, and solve for the energies and wavefunctions of this system. Make sure that your wavefunctions are normalized. Compare your result with that obtained in the text under the approximation $S_{12} = 0$. Would you expect the neglect of overlap approximation to be better for H_2 than for N_2, or *vice versa*? Explain.

7.4 We have introduced the concept of bond order within the context of a homonuclear diatomic molecule. For a heteronuclear diatomic, AB, let the basis states be atomic orbitals $|\phi_A\rangle$ and $|\phi_B\rangle$. Using the neglect of overlap approximation and the Hückel approximation for the Hamiltonian matrix elements, find the bond order of the AB molecule as a function of electron occupation. You may let $\mathscr{H}_{AA} = \mathscr{E}_A$ and $\mathscr{H}_{BB} = \mathscr{E}_B$. Make a schematic plot of your results.

7.5 Consider a molecule that consists of three identical atoms. Given that the bond lengths are identical, there are three possible molecular geometries: linear chain, equilateral triangle, and bent chain with C_{2v} symmetry. The Hamiltonian matrix elements are zero except for those between nearest neighbors, which are $|-\beta|$. For all three configurations, find the eigenvalues and the wavefunctions. What is the most stable configuration as a function of the number of electrons populating the system?

7.6 Consider a molecule A_3 which has point symmetry C_{2v}. Given the tight-binding approximation for the wavefunction,

$$|\psi\rangle = \sum_{j=1}^{3} c_j |\phi(j)\rangle$$
$$= c_1 |\phi(1)\rangle + c_2 |\phi(2)\rangle + c_3 |\phi(3)\rangle \qquad (P7.3)$$

set up the *full* secular equations without *any* approximations. You may assume, however, that there is only one valence orbital on each atom available for bonding in the molecule. This orbital will remain of unspecified nature. Now make the neglect of overlap approximation and assume that all adjacent hopping integrals are β. Furthermore, let the on-site Hamiltonian matrix elements be α. If the second-order hopping integral between atom 1 and atom 3 is of the form $\mathscr{H}_{13} = \mathscr{H}_{31} = \beta \cos(3\theta/2)$, find the angle θ that minimizes the energy as a function of electron occupation.

8

Translational invariance and reciprocal space

We have spent a great deal of effort laying the foundations of group theory and simple models of bonding to prepare us to discuss bonding in the solid state. We have almost all the requisite background now, except that we lack the knowledge of what to do if our object is a crystal and not a molecule. You will recall that the major distinction between crystals and molecules is that the former have translational symmetry. We turn now to the consequences of translational symmetry and how it leads naturally to the concept of reciprocal space.

8.1 Translational symmetry

To make things simple, let us begin with a crystal that has no operations other than E and translation operations. The primitive unit cell has edges $\vec{a}_1$, $\vec{a}_2$, and $\vec{a}_3$, as in Fig. 8.1, and every point in this infinite lattice can be reached by the translation operation

$$\vec{t}_n = n_1\vec{a}_1 + n_2\vec{a}_2 + n_3\vec{a}_3 \tag{8.1}$$

where n_1, n_2, and n_3 are integers. All the atoms in this infinite crystal transform among themselves under all the symmetry operations of the group. This group is the collection of all the translation operations described by $\vec{t}_n$ and the identity, E.

Admittedly, we never really have infinite crystals, so what do we do about the surfaces of the crystal? If the crystal is sufficiently large, then the surface atoms form a negligible part of the total crystal and we can apply *periodic* or *cyclic* boundary conditions. This has the effect of making each direction in the crystal into a circle, with the first atom in that direction of the 'chain' attached to the last atom in the 'chain'. It is quite difficult to visualize this object in three dimensions, but it is certainly clear in one dimension, as

118

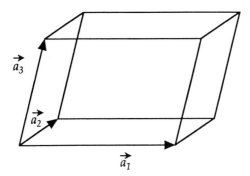

Fig. 8.1. The primitive unit cell of a crystal with no symmetry operations other than E and translation operations.

shown in Fig. 8.2. Our crystal then contains N_1 unit cells in the $\vec{a}_1$ direction, N_2 in the $\vec{a}_2$ direction, *etc.* $n_1, n_2,$ and n_3 are then restricted to take on values $0 \le n_i \le N_i - 1$, so that

$$\vec{t}_{n_1,n_2,n_3} = \vec{t}_{n_1+N_1,n_2,n_3} = \vec{t}_{n_1,n_2+N_2,n_3} = \dots \tag{8.2}$$

Exercise 8.1 Show that the collection of translation vectors forms a group. The group multiplication operation is vector addition.

In addition, since we applied periodic boundary conditions to these lattice translation vectors, this group is cyclic: all of the elements in the group can be generated from one element. Recall also that for cyclic groups, all elements are in a class by themselves. For simplicity, consider a one-dimensional group. The generating element is $A \equiv \{E|\vec{a}_1\}$ and

$$A^{N_1} = E \tag{8.3}$$

A representation of A is the N_1 roots of unity, since these roots have the same multiplication table as the operations

$$\Gamma(\{E|\vec{a}_1\}) = \Gamma(A) = [\exp(-2\pi \imath m)]^{1/N_1} \tag{8.4}$$

where m is any integer.

The order of the translation group is N_1, which is equal to the number of classes and the number of irreducible representations. These conditions put restrictions on m. There are N_1 values of m and it is conventional to pick $0 \le m \le N_1 - 1$.

What are the irreducible representations of the translation group? We know that all the representations must be one-dimensional and there are

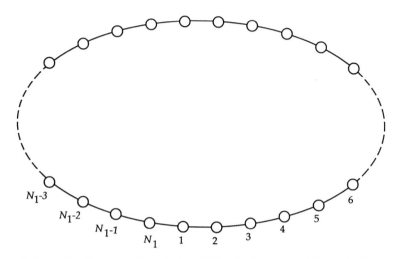

Fig. 8.2. Schematic diagram of a chain of identical atoms with periodic boundary conditions applied. Atom '1' is connected to atom 'N_1'.

N_1 of them. They are listed in Table 8.1. This table is read in precisely the same way as a regular character table: the columns are the various symmetry operations in the group (each one is in its own class) and the rows are the irreducible representations. The only difference is that the last column usually appears as the first column. Note also that we have the totally symmetric irreducible representation. The entries in this table are the characters for the translation group. For example, for the mth irreducible representation, there are some basis functions which, when operated on by the A symmetry operation, result in ϵ^m times the basis function. The rest of the columns are obtained by squaring (A^2), cubing (A^3), etc., the results in the first column. The integers m provide a label for the irreducible representations of the translation group.

Exercise 8.2   Verify the orthonormality of the translation group irreducible representations.

The extension to three dimensions is straightforward,

$$\Gamma(\{E|\vec{t}_n\})_{m_1,m_2,m_3} = \exp(-2\pi\imath m_1 n_1/N_1)\exp(-2\pi\imath m_2 n_2/N_2)$$
$$\times \exp(-2\pi\imath m_3 n_3/N_3) \qquad (8.5)$$

Table 8.1. *Character table of the cyclic translation group of order N*

	A	A^2	$\cdots$	A^{N-1}	$A^N = E \ [\epsilon \equiv e^{-\frac{2\pi i}{N}}]$
$m = 0$	1	1	$\cdots$	1	1
1	ϵ	ϵ^2	$\cdots$	ϵ^{N-1}	$\epsilon^N = 1$
2	ϵ^2	ϵ^4	$\cdots$	ϵ^{2N-2}	$\epsilon^{2N} = 1$
$\vdots$	$\vdots$	$\vdots$		$\vdots$	
$N-1$	ϵ^{N-1}	ϵ^{2N-2}	$\cdots$		$\epsilon^{N^2-1} = 1$

8.2 Reciprocal space

We can give the characters of the translation group a geometric interpretation that turns out to be very fruitful. In real, or direct, space we have the set of primitive vectors $\mathring{a}_1, \mathring{a}_2, \mathring{a}_3$. Define a *reciprocal* lattice vector by

$$\mathring{a}_i \cdot \vec{b}_j = 2\pi \delta_{ij} \tag{8.6}$$

where the reciprocal lattice vectors are given explicitly by

$$\vec{b}_1 = 2\pi \frac{\mathring{a}_2 \times \mathring{a}_3}{\mathring{a}_1 \cdot (\mathring{a}_2 \times \mathring{a}_3)}$$

$$\vec{b}_2 = 2\pi \frac{\mathring{a}_3 \times \mathring{a}_1}{\mathring{a}_1 \cdot (\mathring{a}_2 \times \mathring{a}_3)}$$

$$\vec{b}_3 = 2\pi \frac{\mathring{a}_1 \times \mathring{a}_2}{\mathring{a}_1 \cdot (\mathring{a}_2 \times \mathring{a}_3)} \tag{8.7}$$

Exercise 8.3   Verify that the vectors in Eqs. 8.7 satisfy Eq. 8.6.

A lattice of points is defined by these reciprocal lattice vectors:

$$\vec{K}_n = n_1\vec{b}_1 + n_2\vec{b}_2 + n_3\vec{b}_3 \tag{8.8}$$

where n_1, n_2 and n_3 are integers. Reciprocal space is the space of points defined by the lattice of points $\vec{K}_n$.

In reciprocal space we can define a set of points

$$\vec{k} = p_1\vec{b}_1 + p_2\vec{b}_2 + p_3\vec{b}_3 \tag{8.9}$$

which looks remarkably similar to Eq. 8.8 above, but the p_i's are restricted to values less than 1. Using Eqs. 8.1 and 8.9, we can show that

$$\begin{aligned}
\vec{k} \cdot \vec{t}_n &= (p_1\vec{b}_1 + p_2\vec{b}_2 + p_3\vec{b}_3) \cdot (n_1\mathring{a}_1 + n_2\mathring{a}_2 + n_3\mathring{a}_3) \\
&= 2\pi n_1 p_1 + 2\pi n_2 p_2 + 2\pi n_3 p_3
\end{aligned} \tag{8.10}$$

If we define

$$p_i \equiv \frac{m_i}{N_i} \tag{8.11}$$

and carry over the restriction on m_i ($0 \le m_i \le N_i - 1$), then p_i is restricted to

$$0 \le p_i \le \frac{N_i - 1}{N_i} \tag{8.12}$$

so that p_i varies from $0 \to 1$. We use the label $\vec{k}$ for the irreducible representation of the translation group in place of the label m_i. Comparing Eq. 8.10 and Eq. 8.5, the elements are then replaced by

$$\Gamma(\{E|\vec{t}_n\}) = \exp(-\imath \vec{k} \cdot \vec{t}_n) \tag{8.13}$$

The allowed values of $\vec{k}$ are finite and discrete, but so densely packed in reciprocal space that they may as well form a continuum.

Exercise 8.4   Find the minimum spacing between $\vec{k}$ points.

In actuality, the allowed values of $\vec{k}$ are chosen to be symmetric about $\vec{k} = 0$. The symmetrical cell which bounds the allowed $\vec{k}$ values is known as the first Brillouin zone. We will discuss Brillouin zones in more detail later on.

8.3 Bloch functions and Bloch's theorem

We have found the irreducible representations of the cyclic translation group and have discussed a geometric interpretation for the label of these irreducible representations, $\vec{k}$. It was discussed briefly above that, since the irreducible representations are one-dimensional, the action of a symmetry operation on a basis function was simply to multiply the basis function by the appropriate character. But what are the basis functions for these irreducible representations?

Let us step back a minute and ask a more fundamental question. What does it mean to be a basis function? Let the basis function of the kth irreducible representation be labeled $\phi_{\vec{k}}(\vec{r})$. Then, from the meaning of the translation operation,

$$\{E|\vec{t}_n\}\phi_{\vec{k}}(\vec{r}) = \phi_{\vec{k}}(\vec{r} - \vec{t}_n) \tag{8.14}$$

or, identically, from the definition of a basis function, Eq. 5.36,

$$\{E|\vec{t}_n\}\phi_{\vec{k}}(\vec{r}) = \exp(\imath \vec{k} \cdot \vec{t}_n)\phi_{\vec{k}}(\vec{r}) \tag{8.15}$$

Clearly, the right-hand sides of Eqs. 8.14 and 8.15 must be equal. Try a function of the form

$$\phi_{\vec{k}}(\vec{r}) = \exp(\imath \vec{k} \cdot \vec{r}) u_{\vec{k}}(\vec{r}) \tag{8.16}$$

From Eq. 8.14, we have

$$\phi_{\vec{k}}(\vec{r} - \vec{t}_n) = \exp[\imath \vec{k} \cdot (\vec{r} - \vec{t}_n)] u_{\vec{k}}(\vec{r} - \vec{t}_n) \tag{8.17}$$

and from Eq. 8.15, we have

$$\exp(-\imath \vec{k} \cdot \vec{t}_n) \phi_{\vec{k}}(\vec{r}) = \exp[\imath \vec{k} \cdot (\vec{r} - \vec{t}_n)] u_{\vec{k}}(\vec{r}) \tag{8.18}$$

Therefore,

$$u_{\vec{k}}(\vec{r}) = u_{\vec{k}}(\vec{r} - \vec{t}_n) \tag{8.19}$$

so that the function $u_{\vec{k}}(\vec{r})$ has the translational symmetry of the lattice. The function $\phi_{\vec{k}}(\vec{r})$ is a *Bloch function*. These functions are the basis functions of the kth irreducible representation of the translation group.

Are these Bloch functions eigenfunctions of the Hamiltonian? If so, what are the restrictions on $\mathcal{H}$? We can associate a translation *operator*, $\mathcal{T}_n$, with the translation operation, $\vec{t}_n$. The action of $\mathcal{T}_n$ on any function is described by Eq. 8.14. Further, we can show that $\mathcal{T}_n$ commutes with the kinetic energy part of $\mathcal{H}$,

$$\mathcal{T}_n(-\frac{\hbar^2}{2m}\nabla^2) = (-\frac{\hbar^2}{2m}\nabla^2)\mathcal{T}_n \tag{8.20}$$

From the right-hand side of Eq. 8.20, write

$$(-\frac{\hbar^2}{2m}\nabla^2)\mathcal{T}_n\psi(\vec{r}) = (-\frac{\hbar^2}{2m}\nabla^2)\psi(\vec{r} - \vec{t}_n) \tag{8.21}$$

where $\psi(\vec{r})$ is any function that has the periodicity of the lattice. From the left-hand side of Eq. 8.20, we have

$$\mathcal{T}_n(-\frac{\hbar^2}{2m}\nabla^2)\psi(\vec{r}) \tag{8.22}$$

Comparing Eqs. 8.22 and 8.21, we can see that the requirement that the kinetic energy and the translation operator commute comes down to showing that

$$\nabla^2 = \frac{\partial^2}{\partial x^2} + \frac{\partial^2}{\partial y^2} + \frac{\partial^2}{\partial z^2} + = \frac{\partial^2}{\partial(x - x_n)^2} + \frac{\partial^2}{\partial(y - y_n)^2} + \frac{\partial^2}{\partial(z - z_n)^2} \tag{8.23}$$

which is certainly true for our arbitrary function $\psi(\vec{r})$, which is periodic.

Now, for the translation operator and the potential energy to commute, it must be true that

$$\mathscr{T}_n V(\vec{r}) = V(\vec{r})\mathscr{T}_n \tag{8.24}$$

which requires that

$$V(\vec{r}) = V(\vec{r} - \vec{t}_n) \tag{8.25}$$

Exercise 8.5   Verify Eqs. 8.23 and 8.25.

From our discussion of the parity operator Π (See Problems, Ch. 6) recall that if two operators commute, then they can possess a common set of eigenfunctions. We are now in a position to state *Bloch's theorem.*

Theorem 8.1 *The eigenstates $\phi_{\vec{k}}(\vec{r})$ of the one-particle Hamiltonian, $\mathscr{H} = -(\hbar^2/2m)\nabla^2 + V(\vec{r})$ where $V(\vec{r}) = V(\vec{r} + \vec{t}_n)$, can be chosen to have the form:*

$$\phi_{\vec{k}}(\vec{r}) = e^{i\vec{k}\cdot\vec{r}}u_{\vec{k}}(\vec{r}) \tag{8.26}$$

where $u_{\vec{k}}(\vec{r})$ and $V(\vec{r})$ have the same periodicity.

So this mysterious wavevector $\vec{k}$ that always appeared as if by magic in the wavefunction of an electron in a periodic potential is simply the label of one of the irreducible representations of the translation group. The Bloch wavefunctions are just the basis functions for all the irreducible representations of the translation group. We see that *translational symmetry* is essential for the existence of the wavevector $\vec{k}$. When we come to materials that lack translational symmetry, we will have to adopt an alternative strategy.

8.4 Case study: one-dimensional ring of hydrogen atoms

In this section we will consider a generalization of the H_2 molecule to a very *large* molecule containing N_1 hydrogen atoms, spaced a distance a apart. We apply periodic boundary conditions to this chain by connecting the loose ends so that this system is identical with Fig. 8.2. Although this molecule does not in fact exist (it is energetically favorable for the system to break up into a collection of H_2 molecules), it is an exact model for many cyclic hydrocarbons and a paradigm for many polymers; even selenium atoms can form long chain-like molecules.

Following the discussion in the last section of Ch. 7, we expand the molecular wavefunction in terms of the $1s$ atomic orbital basis states,

$$|\psi\rangle = \sum_{j=1}^{N_1} c_j |\phi_{1s}(j)\rangle \qquad (8.27)$$

Plugging this wavefunction into the Schrödinger equation, we obtain an $N_1 \times N_1$ matrix for the equations,

$$\begin{pmatrix} \mathcal{H}_{11} - \mathcal{E}S_{11} & \mathcal{H}_{12} - \mathcal{E}S_{12} & \cdots & \mathcal{H}_{1N_1} - \mathcal{E}S_{1N_1} \\ \mathcal{H}_{21} - \mathcal{E}S_{21} & \mathcal{H}_{22} - \mathcal{E}S_{22} & \cdots & \cdots \\ \vdots & \vdots & \vdots & \vdots \\ \mathcal{H}_{N_11} - \mathcal{E}S_{N_11} & \cdots & \cdots & \cdots \end{pmatrix} \begin{pmatrix} c_1 \\ c_2 \\ \vdots \\ c_{N_1} \end{pmatrix} = 0$$

As we pointed out in Ch. 7, this is a messy problem to solve, even if it does have a finite number (N_1) of eigenvalues and eigenvectors.

To simplify the problem somewhat, we invoke the neglect of overlap approximation and apply the Hückel approximation. In addition, since all sites in the molecule have identical atoms, we let $\mathcal{H}_{11} = \mathcal{H}_{22} = \ldots \equiv \alpha$, where α is the traditional shorthand for the 'on-site' Hamiltonian matrix elements. The off-diagonal matrix elements, the hopping integrals, are denoted β. We thus have

$$\begin{pmatrix} \alpha - \mathcal{E} & \beta & 0 & \cdots & \beta \\ \beta & \alpha - \mathcal{E} & \beta & \cdots & 0 \\ 0 & \beta & \alpha - \mathcal{E} & \cdots & 0 \\ \vdots & \vdots & \vdots & \vdots & \vdots \\ \beta & 0 & 0 & \cdots & \alpha - \mathcal{E} \end{pmatrix} \begin{pmatrix} c_1 \\ c_2 \\ c_3 \\ \vdots \\ c_{N_1} \end{pmatrix} = 0$$

Notice the very regular structure to this matrix. The diagonal elements are always $\alpha - \mathcal{E}$, while the sub- and super-diagonal elements are β. The upper rightmost element and lower leftmost element are also β. This structure reflects the periodic boundary conditions. All other elements are zero.

At this point it is standard to divide through by β (we are sure that it is nonzero), and for $x \equiv (\alpha - \mathcal{E})/\beta$, we have

$$\begin{pmatrix} x & 1 & \cdots & \cdots & 1 \\ 1 & x & 1 & \cdots & 0 \\ 0 & 1 & x & 1 & 0 \\ \vdots & \vdots & \vdots & \vdots & \vdots \\ 1 & 0 & 0 & 1 & x \end{pmatrix} \begin{pmatrix} c_1 \\ c_2 \\ c_3 \\ \vdots \\ c_{N_1} \end{pmatrix} = 0$$

All equations are of the identical form,

$$c_{j-1} + x c_j + c_{j+1} = 0 \qquad (8.28)$$

if we define $c_0 = c_{N+1} \equiv 0$. Quantum chemists take the approach of using the symmetry properties of our large molecule to factor the determinant of the coefficients into smaller expressions and arrive at the eigenvalues. We can use this same philosophy as well, but keep the approach more general, since we have a very large system.

Firstly, note that our H atom ring has cyclic symmetry, so its point group has the same structure as the translation group. An educated guess for the solutions to Eq. 8.28 is a linear combination of exponential functions containing the magnitude of k and the atom index j,

$$c_j = A e^{ikja} + B e^{-ikja} \tag{8.29}$$

The definition of the coefficient c_0 and a bit of tedious algebra gives us

$$c_j = A(e^{ikja} - e^{-ikja}) \tag{8.30}$$

and

$$x = -2\cos(ka) \tag{8.31}$$

Exercise 8.6   Verify Eqs. 8.30 and 8.31.

Now, we already know the restrictions on the values of k. In one real space dimension, we only have one reciprocal space dimension, given by

$$b = \frac{2\pi}{a} \tag{8.32}$$

from Eq. 8.6. Therefore, from Eqs. 8.9 and 8.12, k can take on values

$$k = 0, \frac{2\pi}{a}\frac{1}{N_1}, \ldots, \frac{2\pi}{a}\frac{N_1 - 1}{N_1} \tag{8.33}$$

Or, choosing k to be symmetric about $k = 0$, the wavevector lies in the bounds

$$-\frac{\pi}{a} \le k \le \frac{\pi}{a} \tag{8.34}$$

Remember that there are N_1 distinct values of k, and that they are distributed uniformly over an interval of width $2\pi/a$. For $a = 1.5$ Å, a rough estimate of the spacing between H atoms, then each distinct value of k is spaced $(2\pi/a)(1/(N_1 - 1))$ apart, which gets very small indeed as N_1 gets large.

Exercise 8.7   For several values of N_1, find the spacing between neighboring $\vec{k}$ points.

Finally, from Eqs. 8.27 and 8.30, we can find the value of A by invoking normalization,

$$
\begin{aligned}
\langle \psi | \psi \rangle &= \sum_{i,j} c_i^* c_j \langle \phi_{1s}(i) | \phi_{1s}(j) \rangle \\
&= \sum_{i,j} c_i^* c_j \delta_{ij} \\
&= \sum_j c_j^* c_j \\
&= \sum_j^{N_1} A^* A (2\sin(kja))^2 \qquad (8.35)
\end{aligned}
$$

so that

$$
A = \frac{1}{2\sqrt{N_1}} \qquad (8.36)
$$

Exercise 8.8   Verify the normalization condition.

Bringing all these results together, the eigenvalues of the H ring molecule are:

$$
\mathcal{E} = \alpha + 2\beta \cos(ka) \qquad (8.37)
$$

and the wavefunctions are:

$$
| \psi \rangle = \frac{1}{\sqrt{N_1}} \sum_{j=1}^{N_1} \cos(kja) | \phi_{1s}(j) \rangle \qquad (8.38)
$$

Exercise 8.9   Is this wavefunction a Bloch function? Why?

Recall that for basis functions that are s atomic orbitals, $\beta < 0$. We plot in Fig. 8.3 the energy as a function of k for unknown values of the parameters α and β. This is our first example of a *band diagram*. The name arises because we no longer have the discrete energy states as we did in molecules, but rather we have a (nearly) continuous *band* of energies. As N_1 gets large, there are more and more allowed values of k which are very closely spaced. The *band width* is the magnitude of the difference between the maximum and minimum energies. In this case, the band width is $4|\beta|$, with the maximum energy at $k = \pm(\pi/a)$ and the minimum energy at $k = 0$. If we have a strong interaction between neighboring atoms, $|\beta|$ is large and the band width is also large. On the other hand, for weakly interacting neighboring atoms, $|\beta|$ is small and we have narrow bands.

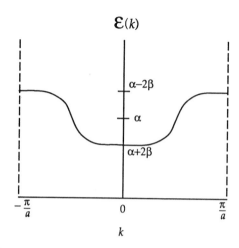

Fig. 8.3.   The band structure of the one-dimensional ring of H atoms. α and β are unknown parameters.

What about the wavefunctions? At $k=0$,

$$|\psi\rangle = \frac{1}{\sqrt{N_1}} \sum_{j=1}^{N_1} |\phi_{1s}(j)\rangle \tag{8.39}$$

so that all atomic orbitals add with the same phase, as shown in (a) of Fig. 8.4. This is clearly a bonding state. At $k = \pi/a$,

$$|\psi\rangle = \frac{1}{\sqrt{N_1}} \sum_{j=1}^{N_1} (-1)^j |\phi_{1s}(j)\rangle \tag{8.40}$$

so that the atomic orbitals add with alternating phases, as shown in (b) of Fig. 8.4. This is clearly an antibonding state. The fact that the $k = 0$ state is lower in energy than the $k = \pi/a$ state is consistent with the total wavefunction, since the bonding state is lower in energy than the antibonding state. Other values of wavevector k give a wavefunction that is neither completely bonding nor completely antibonding. The energy of any intermediate state must therefore fall between these two extremes. What we want to emphasize at this point is the correspondence between the energy of an electron state in a band diagram and the real space picture of the wavefunction.

Essentially no new ideas have emerged in our treatment of a very large hydrogen-based molecule as a one-dimensional crystal. The concepts that

(a) Bonding state

$k = 0$

(b) Antibonding state

$k = \pm\frac{\pi}{a}$

Fig. 8.4. Schematic diagram of the real-space pictures of the (a) bonding ($k=0$) and (b) antibonding ($k = \pm\pi/a$) part of the wavefunction. The shaded s-type orbitals have opposite phase from the unshaded orbitals. The spacing between atoms is a.

we discovered in the diatomic molecule are still qualitatively valid; we just have many more states now and they constitute a *band*.

Problems

8.1   Consider a closed loop of atoms as in Fig. 8.2. However, assume that the s atomic orbitals are very low in energy, so that they can be considered as 'core' orbitals and do not participate in bonding. The highest-energy occupied orbitals turn out to be p_z orbitals, where the z axis is perpendicular to the ring. Find the band structure for this molecule. How does it differ, if at all, from the case of the s atomic orbital basis?

8.2   (a) Show that the irreducible representations of the translation group are orthonormal (*i.e.*, that they obey GOT).

   (b) In classical mechanics, the kinetic energy of a particle is given by $p^2/2m$, where p is the momentum. The kinetic energy operator in quantum mechanics is $-(\hbar^2/2m)\nabla^2$, which implies that the momentum operator is $\mathbf{p} = (\hbar/\imath)\nabla$. Show that the Bloch functions cannot be eigenfunctions of the momentum operator. This result casts some doubt on the conclusion that many like to draw that $\hbar\vec{k}$ corresponds to the electron momentum. For the present, then, we must be content to view $\vec{k}$ as the label of the irreducible representations of the translation group.

8.3   Consider the ring of H atoms discussed at length in this chapter. The band structure $\mathscr{E}(k)$ we derived is

$$\mathscr{E}(k) = \alpha + 2\beta \cos(ka) \qquad \text{(P8.1)}$$

where α is the on-site Hamiltonian matrix element, β is the hopping matrix element between adjacent atoms, and k is the wavevector which has limits $-(\pi/a)$ and (π/a) for an atomic spacing equal to a. All other Hamiltonian matrix elements are zero.

(a) The group velocity of the electron in the eigenstate ψ_k is defined by

$$v_k = \frac{1}{m}\langle\psi_k|p|\psi_k\rangle \tag{P8.2}$$

where

$$p = \frac{\hbar}{\iota}\frac{d}{dx} \tag{P8.3}$$

is the momentum operator and m is the electron mass. Using a Bloch wavefunction and the time-*independent* Schrödinger equation, show that the group velocity is

$$v_k(x) = \frac{1}{\hbar}\frac{d\mathscr{E}(k)}{dk} \tag{P8.4}$$

Hint: Insert the Bloch wavefunction into the Schrödinger equation with $\mathscr{H} = -(\hbar^2/2m)\nabla^2 + V(x)$. Differentiate this expression with respect to k. Compare this result with $(1/m)\langle\psi_k|p|\psi_k\rangle$.

(b) Show that eigenstates labeled by k and $-k$ have equal and opposite velocities. Show that the group velocity is zero at the Brillouin zone boundaries.

(c) We can use the time-*dependent* Schrödinger equation to get more insight into the meaning of the group velocity. A more generic expression for the wavefunctions of the H atom ring is

$$|\psi_k\rangle = \frac{1}{\sqrt{N}}\sum_{j=1}^{N}c_j^{(k)}|j\rangle \tag{P8.5}$$

Recall that in this expression j labels the atomic position and k is a label for an irreducible representation of the translation group. In the kth eigenstate the amount of electronic charge associated with atom j at time t is $-e|c_j^{(k)}(t)|^2$, where e is the electronic charge. This quantity changes with time, and the current flowing onto or off of site j in the kth eigenstate is given by the time derivative of the charge. Using the time-dependent Schrödinger equation, find a general expression for the current flowing onto or off of a site.

Using the appropriate time dependence and Bloch functions,

$$c_j^{(k)}(t) = e^{ijka}\exp\left(-\frac{\iota\mathscr{E}t}{\hbar}\right) \tag{P8.6}$$

show that the current flowing onto or off a site is proportional to the group velocity.

8.4   Consider an infinite linear chain of alternating A and B atoms with periodic boundary conditions. The on-site Hamiltonian matrix elements are $\mathscr{E}_A$ on A atoms and $\mathscr{E}_B$ on B atoms. The only other nonzero matrix elements are the first-order hopping integrals β. The A–B bond length is a. Calculate and sketch the band structure and show that there are *two* bands and that the Brillouin zone may be defined to lie between $-(\pi/2a)$ and $(\pi/2a)$. Draw schematic pictures of the wavefunctions for the two bands and discuss the bonding nature of the two bands.

8.5   Exercise 8.1.

8.6   Exercise 8.5.

8.7   Exercise 8.8.

8.8   Exercise 8.9.

9

Characterization of electrons in a solid

9.1 Introduction

In the last chapter we made an important leap in attempting to understand the electronic structure of materials. Using the translational symmetry of a fictitious crystal that was infinite in extent, we wrote down expressions for the energy and wavefunctions. We saw that both of these quantities relied on a vector $\vec{k}$, the wavevector. Even though we generated real space pictures of the crystal wavefunctions, the energy was presented as a *band structure* of energy bands *vs.* $|\vec{k}|$, where $|\vec{k}|$ was restricted to a certain interval, the first Brillouin zone. The wavevector arose strictly from the translational symmetry of the crystal and, within the context of group theory, k was just a label for one of the irreducible representations of the cyclic translation group. The Bloch functions were just the functions that generated these irreducible representations.

Bloch's theorem, stating that the wavefunctions of a system with a periodic potential can be written as the product of an exponential and a function that possessed the symmetry of the crystal, was a consequence of the commutativity of $\mathscr{H}$ and $\mathscr{T}_n$, the translation operator. Translational symmetry clearly forms a cornerstone of understanding electronic structure of crystalline materials.

One additional point that bears mentioning is that the expression for the band energy

$$\mathscr{E}(k) = \alpha + 2\beta \cos(ka) \tag{9.1}$$

is just a truncated Fourier transform of the Hamiltonian matrix elements. Remember that we have worked explicitly in the Hückel approximation, so that only matrix elements connecting adjacent atoms are nonzero. What if we allow matrix elements between atoms farther apart to be nonzero? How does the expression for the energy change? Writing a few terms in the

132

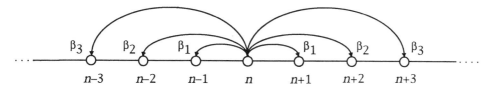

Fig. 9.1. A schematic diagram of the atomic interactions that give rise to the various hopping integrals.

Fourier transform of the Hamiltonian matrix elements,

$$
\begin{aligned}
\mathcal{E}(k) &= \sum_{j=1}^{N} e^{\iota(j-n)ka} \langle n|\mathcal{H}|j\rangle \\
&= \mathcal{H}_{nn} + \mathcal{H}_{n,n+1}(e^{\iota ka}) + \mathcal{H}_{n-1,n}(e^{-\iota ka}) \\
&\quad + \mathcal{H}_{n,n+2}(e^{2\iota ka}) + \mathcal{H}_{n-2,n}(e^{-2\iota ka}) \\
&\quad + \mathcal{H}_{n,n+3}(e^{3\iota ka}) + \mathcal{H}_{n-3,n}(e^{-3\iota ka}) + \cdots
\end{aligned}
\tag{9.2}
$$

where in the Fourier transform n is some atom site between 1 and N. $\mathcal{H}_{nn}$ is just the on-site energy, α. Furthermore, because of translational symmetry, the matrix elements between atom n and *both* its neighbors, $n+1$ and $n-1$, are identical. This allows us to combine higher terms in pairs so that

$$
\mathcal{E}(k) = \alpha + 2\mathcal{H}_{n,n+1}\cos(ka) + 2\mathcal{H}_{n,n+2}\cos(2ka) + 2\mathcal{H}_{n,n+3}\cos(3ka) + \cdots \tag{9.3}
$$

The hopping integrals can be denoted as β_1, β_2, β_3,..., where the subscript indicates first neighbor, second neighbor, *etc.* These hopping integrals are also sketched schematically in Fig. 9.1. Finally, then,

$$
\mathcal{E}(k) = \alpha + 2\beta_1\cos(ka) + 2\beta_2\cos(2ka) + 2\beta_3\cos(3ka) + \cdots \tag{9.4}
$$

The first two terms give our original result from Ch. 8, which is redrawn in Fig. 9.2 as the heavy solid line. The other terms add *higher-frequency* components to this band structure, so that a more accurate band structure has a great deal more curvature in it. The *amplitude* of the higher-order terms diminishes as the separation between atoms grows, however. Some argue that Hamiltonian matrix elements scale like d^{-2}, where d is the separation between atoms. Therefore, our higher-order terms are perturbations on the basic shape derived under the Hückel approximation.

   In this chapter we will introduce a number of concepts that allow us to have less reliance on $\vec{k}$ and translational symmetry. We hope to develop a picture of bonding that is rooted firmly in *real* space, as opposed to

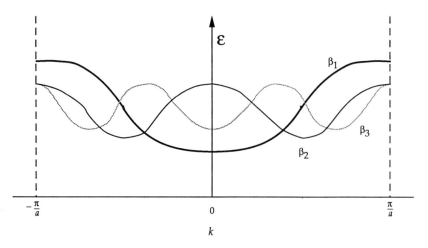

Fig. 9.2. Band structure for the periodic ring of H atoms. The heavy solid line represents the contribution from nearest-neighbor interactions. The lighter solid line is the contribution from second-neighbor interactions (β_2) and the dotted-patterned line is the contribution from third-neighbor interactions (β_3).

reciprocal space, for not only do we inhabit real space, but also we cannot rely on translational symmetry to study the wide range of materials that are the domain of modern materials science.

9.2 The density matrix

We presented some of the wavefunctions for the simple H ring at the end of Ch. 8. There we found that, for $k=0$, the phases of all the $1s$ atomic orbitals were identical and, hence, this looked like an expanded version of the bonding state in H_2. We also found that for $k = \pi/a$, the 'zone boundary', the phases of the $1s$ atomic orbitals alternated, making this state antibonding. If we choose $0 < k < \pi/a$, we find a more complicated wavefunction, in that some orbitals have positive phase, others have negative phase, and still others have an imaginary coefficient. Wavefunctions for $k = 0$, $k = \pi/5a$, and $k = \pi/a$ are reproduced in Fig. 9.3.

The coefficients, c_j, in the crystal wavefunction expansion serve to modulate the amplitude and sign of the atomic basis orbitals. Figure 9.3 shows a sketch of both the atomic basis orbitals and the coefficients. The latter form an envelope around the former. One important thing to point out is that since our potential is a periodic, even function, $\mathcal{H}$ commutes with Π, the parity operator. The wavefunctions must then have a definite parity, as

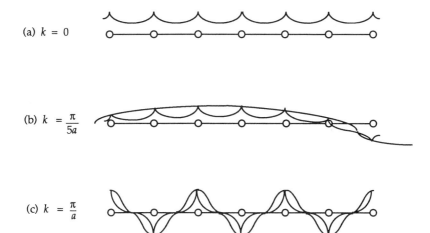

(a) $k = 0$

(b) $k = \dfrac{\pi}{5a}$

(c) $k = \dfrac{\pi}{a}$

Fig. 9.3. Schematic diagrams of the real part of the wavefunction for the periodic H ring at (a) $k = 0$, (b) $k = \pi/5a$, and (c) $k = \pi/a$. For (b) and (c), the modulation from the coefficients is also shown.

can be verified by visual inspection of Fig. 9.3. The most important point about the wavefunction, however, is that it tells us where the electrons are most likely to be found. Recall that we used plots of the charge density in a graphite sheet and in crystalline Si as a motivation and justification for using the tight-binding approximation. However, what was plotted in Figs. 6.2 (graphite sheet) and 6.3 (cross-section of Si) was the *total* charge density, *i.e.*, the *sum* of the probability over all *occupied* states. The charge density is proportional to $\sum_{occ} |\psi(x)|^2 = \sum_{occ} \langle x|\psi \rangle \langle \psi|x \rangle$, where the sum runs over all occupied states and the wavefunctions have been expressed in terms of a real space basis set. We have replotted the wavefunctions of Fig. 9.3 as charge densities in Fig. 9.4.

The key to determining the total charge density is twofold: firstly, we must figure out which states are occupied, and, secondly, we must figure out how to add all these charge densities together. How do we determine whether a given state is occupied? We invoke the Pauli exclusion principle, which says that no two electrons can have exactly the same set of quantum numbers. For our more practical application, this means that we can populate energy eigenstates with only two electrons of opposing spin. Two other empirical rules from molecular chemistry are useful. The *aufbau* principle, from the German 'to build up', simply means that we begin putting electrons into the lowest-energy states and then into those with successively higher and higher energies. Finally, Hund's rules say that for an incompletely occupied

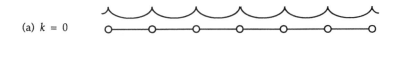

(a) $k = 0$

(b) $k = \dfrac{\pi}{5a}$

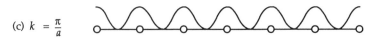

(c) $k = \dfrac{\pi}{a}$

Fig. 9.4. The same wavefunctions as plotted in Fig. 9.3, but now plotted as charge densities.

set of degenerate levels, the states are to be occupied so that the spin is maximized. This means, for example, in a system with cubic site symmetry, the p_x, p_y, and p_z atomic orbitals are degenerate (see the character table for the T point group). Therefore, a molecular wavefunction that is a linear combination of the three p orbitals contains one electron from each atomic orbital until we reach an occupation of four electrons. The fourth, fifth, and sixth electrons by necessity need to be paired. We have implicitly assumed that the temperature is 0 K, a point to which we will return later.

Filling bands in a solid follows precisely the same principles. However, as we will discover, sometimes bands can be fractionally occupied, with the fraction differing from $\frac{1}{2}$. An important point to recognize is that in a solid there are N k points in the Brillouin zone. We can fill each k point with two electrons, which means that our simple one-dimensional band can accommodate a total of $2N$ electrons. The energy which separates the filled states from the unfilled states is called the Fermi energy and it is denoted $\mathscr{E}_F$. There is a corresponding Fermi wavevector, k_F,

$$\mathscr{E}_F = \alpha + 2\beta \cos(k_F a) \tag{9.5}$$

If the band is full, then, from Fig. 9.5, we see that $k_F = \pi/a$, at the zone boundary. For anything less than a completely filled band, k_F is smaller than π/a.

Now it is clear, however, how to obtain the *total* charge density; all we need do is sum over states with wavevectors from $k = -k_F$ to $k = k_F$. The exact expression for the charge density is the probability density multiplied

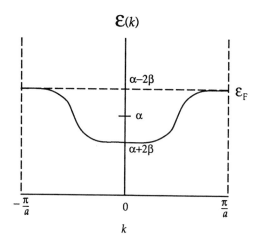

Fig. 9.5.   The band structure of the periodic H ring with the Fermi energy $\mathscr{E}_F$ for a completely occupied band indicated. The corresponding Fermi wavevector is $k_F = \pi/a$.

by $(-2e)$, where the factor of 2 appears because each state can accommodate two electrons and the factor of $(-e)$ appears in accord with the convention that the electron charge $q = -e$.

$$\rho(x) = -2e \sum_{\text{occ}} \langle x|\psi\rangle\langle\psi|x\rangle \tag{9.6}$$

Here the notation $\langle\psi|x\rangle$ means the wavefunction ψ is expressed in a real-space basis. Since we have argued that k is a nearly continuous variable, we can replace the sum over k by an integral,

$$\rho(x) = -2e \frac{Na}{2\pi} \int_{-k_F}^{k_F} dk \, \langle x|\psi\rangle\langle\psi|x\rangle \tag{9.7}$$

The factor $Na/2\pi$ appears from changing the sum into an integral.

Exercise 9.1   Show this last step. Hint: The quantity $Na/2\pi$ is the inverse of the number of states that are in a unit distance along the k axis.

Now, using the Bloch-type wavefunction from Ch. 8,

$$|\psi\rangle = \frac{1}{\sqrt{N}} \sum_{j=1}^{N} e^{ijka}|\phi_j\rangle \tag{9.8}$$

we have,

$$
\begin{aligned}
\rho(x) &= -2e\frac{Na}{2\pi}\frac{1}{N}\sum_{j=1}^{N}\sum_{l=1}^{N}\int_{-k_F}^{k_F} dk\, e^{\imath(j-l)ka}\langle x|\phi_j\rangle\langle\phi_l|x\rangle \\
&= -e\sum_{j=1}^{N}\sum_{l=1}^{N}\left(\frac{a}{\pi}\int_{-k_F}^{k_F} dk\, e^{\imath(j-l)ka}\right)\langle x|\phi_j\rangle\langle\phi_l|x\rangle \\
&= -e\sum_{j=1}^{N}\sum_{l=1}^{N}\rho_{lj}\langle x|\phi_j\rangle\langle\phi_l|x\rangle
\end{aligned}
\tag{9.9}
$$

and the *density matrix elements*, ρ_{lj}, are defined as

$$
\rho_{lj} \equiv \frac{a}{\pi}\left(\int_{-k_F}^{k_F} dk\, e^{\imath(j-l)ka}\right)
\tag{9.10}
$$

The $N \times N$ matrix that contains the elements ρ_{lj} is called the *density matrix*. The diagonal components are just the 'on-site' charge density. The off-diagonal elements, however, are related to the charge density between atoms, or, as we introduced in Ch. 7, the bond order.

Let us look more closely at the density matrix elements. Take the diagonal components first,

$$
\rho_{ll} = \frac{a}{\pi}\int_{-k_F}^{k_F} dk = \frac{2ak_F}{\pi}
\tag{9.11}
$$

What does this mean? First of all, ρ_{ll} does not depend on which site we examine, which is encouraging because all sites are equivalent. Suppose $k_F=0$, so that there are no electrons in the system; Eq. 9.11 indicates that the charge density associated with each atom is zero. On the other hand, if $k_F = \pi/a$ at the zone boundary, we know that the band is full and $\rho_{ll}=2$, which is what we expect. Finally, the charge on each atom varies linearly as k_F moves from the zone center to the zone boundary, which is consistent with our band structure in Fig. 9.5.

Let us turn now to the off-diagonal part, where $l \neq j$. Then,

$$
\begin{aligned}
\rho_{lj} &= \frac{a}{\pi}\int_{-k_F}^{k_F} dk\, e^{\imath(j-l)ka} \\
&= \frac{a}{\pi}\left[\frac{e^{\imath(j-l)k_F a} - e^{-\imath(j-l)k_F a}}{\imath(j-l)a}\right] \\
&= \frac{2}{\pi}\frac{\sin[(j-l)k_F a]}{(j-l)}
\end{aligned}
\tag{9.12}
$$

Equation 9.12 is similar to the expression for the amplitude of light through

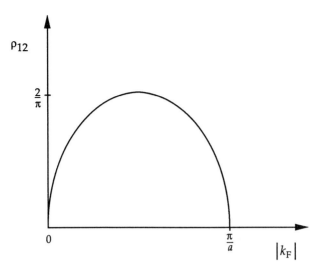

Fig. 9.6. The density matrix element ρ_{12} plotted as a function of band filling. For $k = 0$, the band is completely empty, while for $k = \pi/a$, the band is completely full. The density matrix element is a measure of the bond order.

a single slit. Figure 9.6 is a schematic of Eq. 9.12 for $|j-l| = 1$. This could be $l = 1, j = 2$ or any other adjacent combination. ρ_{lj} has the same translational symmetry as the ring. When the band is empty at $k_F = 0$, clearly there are no electrons to share in bonds. This is reflected in Eq. 9.12 and Fig. 9.6. What happens as we start filling the band, however, is very interesting. Recall that for $k = k_F = \pi/a$, the zone boundary, the wavefunction is completely antibonding in character. This means that if we fully occupy a band, the completely bonding part (the bottom of the band) is occupied as equally as the completely antibonding part (the top of the band). There is no net bonding! But this is precisely the same thing that we saw for the bond order in the H_2 diatomic molecule (see Exercise 7.8). We had, in that case, a bonding and an antibonding eigenstate and the bond order varied with occupation exactly the same way that Fig. 9.6 does.

What about larger separations between atom j and atom l? From Eq. 9.12, as $|j - l|$ increases, the density matrix element decreases, so that the bond order decreases. Furthermore, we get higher-frequency oscillations in the numerator of Eq. 9.12, which simply adds more wiggles on Fig. 9.6.

The insight that the off-diagonal density matrix elements are related to the bond order is very important. We can already envision one simple application. Suppose we wanted to decrease the bond length in our H atom ring. How might we accomplish this? The answer is present in Fig. 9.6 –

we just need to remove electrons from the (fully occupied) band. This will cause the bond order to increase and, hence, the bond length to decrease. Manipulating the properties of solids may not be so mysterious as we thought!

9.3 The density of states

The density matrix provides powerful insight into locating the electrons in a solid: are they confined to the region between neighboring atoms, as in Si or diamond, or are they more delocalized, as in a typical metal like Cu or Al? The information provided by the density matrix is important when we come to looking for *bonds* in a material and, hence, will guide us in understanding structure. However, the energetics of our system are still not completely obvious. The band structure for the one-dimensional ring of H atoms is simple enough. But what happens when we move on to two- and three-dimensional systems? The Brillouin zone will no longer be a simple line. How do we even begin to visualize, much less interpret, the bands in a complicated three-dimensional crystal? Worse still, what happens if our material lacks long-range order, so that the concept of reciprocal space is lost? Fortunately, examining the distribution of states as a function of energy eliminates reliance on k and, as we will see, allows us to some extent to bypass complicated and computationally demanding electronic structure calculations.

The distribution of states within an energy interval from $\mathscr{E}$ to $\mathscr{E} + d\mathscr{E}$ is called the *density of states* (DOS) and is expressed as

$$D(\mathscr{E})d\mathscr{E} = \text{number of states per unit volume between } \mathscr{E} \text{ and } \mathscr{E} + d\mathscr{E} \quad (9.13)$$

Let us construct the DOS for the periodic ring of H atoms using simple arguments based on the band structure, before we present a formal expression. The band structure is replotted on the top of Fig. 9.7. Take an imaginary yardstick of length $d\mathscr{E}$, which will be a tiny fraction of the entire band width. Now place that yardstick close to the bottom of the band, parallel to the energy axis. Because the band is relatively flat in this region, we will count a large number of states in the interval $d\mathscr{E}$ with our yardstick. Suppose now, however, we move the yardstick to a position along the ordinate axis that corresponds to halfway up the band. The greater curvature of the band means that we count fewer states in the *same interval* $d\mathscr{E}$. The top of the band is identical with the bottom of the band. If we think of our yardstick as a bin and the number of states measured with it as the occupation of each bin, then making a histogram plot results in the figure on the bottom

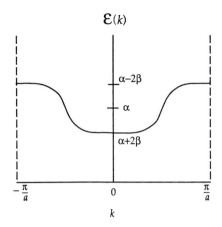

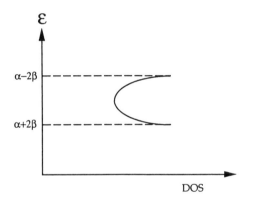

Fig. 9.7. Plot of the band structure of the periodic H ring (top) with a plot of the electron density of states (DOS) (bottom) for the same model.

of Fig. 9.7. The width of the DOS in energy is exactly the same as the band width and the asymptotic approach of the DOS to infinity at the band edges reflects the fact that the slope of $\mathscr{E}(k)$ vanishes as $k = 0, \pm\pi/a$.

We can now write down a more rigorous expression for the DOS. Take the total number of states per unit volume up to an energy $\mathscr{E}$ as S. Then

$$D(\mathscr{E}) = \frac{dS}{d\mathscr{E}} \tag{9.14}$$

or, since $\mathscr{E} = \mathscr{E}(k)$,

$$D(\mathscr{E}) = \frac{dS}{dk} \left| \frac{dk}{d\mathscr{E}} \right|$$

$$= \frac{\frac{dS}{dk}}{\left|\frac{d\mathscr{E}}{dk}\right|} \tag{9.15}$$

where the absolute value is used, since the slope of $\mathscr{E}(k)$ can be negative, but the density of states must clearly always be positive. The numerator, dS/dk, is the density of states in k space, which from above is $Na/2\pi$. Note the inverse dependence of the density of states on the slope of the band structure, as we found from heuristic arguments.

Using Eq. 9.1 for the band structure of the H atom ring, we find

$$\begin{aligned} D(\mathscr{E}) &= 2\frac{Na}{2\pi}\frac{1}{2\beta a \sin(ka)} \\ &= \frac{N}{\pi}\frac{1}{\sqrt{4\beta^2 - (\mathscr{E} - \alpha)^2}} \end{aligned} \tag{9.16}$$

where the second expression follows from inserting Eq. 9.1 and using the trigonometric identity $\sin^2\theta + \cos^2\theta = 1$. This form is preferred, since then the DOS is an explicit expression of energy. There is an additional factor of 2 in Eq. 9.16 because there are two states for every value of $\mathscr{E}$, one at k, another at $-k$. A quick sketch of this function reveals a plot identical with the bottom of Fig. 9.7.

What does the DOS do for us? Simply put, it counts states. If we were to integrate the DOS from the zero of energy (a concept to which we will return shortly) up to the Fermi energy, the result is the total number of occupied states per unit volume. If we multiply this by 2, we get the electron density of the system.

Exercise 9.2   Using Eq. 9.16, verify this last statement for several different values of $\mathscr{E}_F$.

Essentially, what the DOS plots is the distribution of electrons in energy. In addition, the DOS is an average of the bands over k space, which will become more useful as we study more complicated systems.

Do we really need the band structure to construct the DOS *schematically*? The answer is no, provided that we know: (1) the atoms involved in bonding; (2) their approximate ionization potentials and electronegativities; and (3) a rough estimate of the hopping matrix element.

As an example, consider an infinite ring of identical atoms with both s and p valence orbitals. Since all the atoms are the same, the only influence of the ionization potential is through the energy difference between the s and p atomic orbitals. The ionization potential is lower for the p orbitals. How do we know this? Recall that for the H *atom*, the energy eigenvalues depended only on the principal quantum number. However, in any environment that is not spherically symmetric, the degeneracies of the eigenvalues are dictated by the dimensionality of the irreducible representations. Because the s, p, d, etc., orbitals have different symmetries, they will have different energies. Furthermore, because of the different radial dependences for the atomic orbitals of different angular momenta, the s atomic orbitals are lower in energy than the p atomic orbitals, and so on for a given principal quantum number. If we assume that there is no interaction between the s and p atomic orbitals, then the DOS for the s orbital band is identical with Fig. 9.7. The DOS for the various p orbitals is quite different, however. Choosing the z axis as the bond axis, the p_z orbital DOS has a similar shape to the s contribution, but it is higher in energy (different ionization energy) and has a smaller band width (less overlap). The p_x and p_y orbitals give rise to π-type bonds and have an even smaller DOS width. All of these curves are collected in Fig. 9.8.

Exercise 9.3   Why is the band width for the p_x- and p_y-derived bands smaller than the p_z-derived band?

Exercise 9.4   What would the DOS look like for an periodic ring containing atoms with d valence orbitals?

But are we really better off knowing the DOS? If we bypassed solving the Schrödinger equation, then most assuredly so. However, we would still like to know whether, and where, these states are localized. This leads us to the idea of the *local* DOS, in which we project out of the *total* DOS only that part which is of interest to us. For example, to return to the DOS for the H ring, note that there is a factor of N in Eq. 9.16. But there are N identical atoms in the ring! That means that the density of states associated with *each* atom, the local DOS, is just

$$d(\mathscr{E}) = \frac{D(\mathscr{E})}{N}$$

$$= \frac{1}{\pi} \frac{1}{\sqrt{4\beta^2 - (\mathscr{E} - \alpha)^2}} \tag{9.17}$$

The proper notation for the *local* DOS is a lower case $d(\mathscr{E})$.

Consider a general tight-binding wavefunction,

$$|\psi\rangle = \sum_j \langle \phi_j|\psi\rangle|\phi_j\rangle$$

$$= \langle \phi_1|\psi\rangle|\phi_1\rangle + \langle \phi_2|\psi\rangle|\phi_2\rangle + \cdots + \langle \phi_N|\psi\rangle|\phi_N\rangle \tag{9.18}$$

If the expansion coefficient $\langle \phi_j|\psi\rangle = 0$, then atom j does not contribute

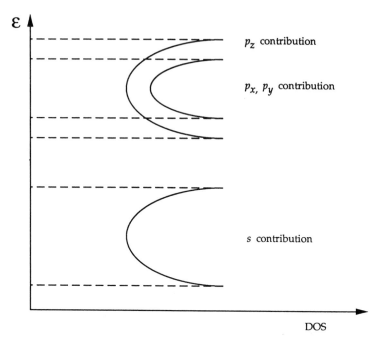

Fig. 9.8. A schematic drawing of the electron DOS for a system of identical atoms with valence *s* and *p* orbitals. It is assumed that the *s* and *p* bands do not interact.

to this eigenstate. However, if $\langle \phi_n | \psi \rangle \neq 0$ for atom n, then $|\langle \phi_n | \psi \rangle|^2$ is the weighting factor for atom n's contribution to the eigenstate $|\psi\rangle$. If the eigenvalue of this state is $\mathscr{E}_k$, then the state contributes $\left(\frac{dS}{d\mathscr{E}} \right)_{\mathscr{E}=\mathscr{E}_k}$ to the *total* DOS. The local DOS is thus this contribution multiplied by the weighting factor, summed over all eigenvalues,

$$ d_n(\mathscr{E}) = \sum_{\text{all eigenvalues } \mathscr{E}_k} \left(\frac{dS}{d\mathscr{E}} \right)_{\mathscr{E}=\mathscr{E}_k} |\langle \phi_n | \psi \rangle|^2 \qquad (9.19) $$

The key to understanding Eq. 9.19 is our use of the word 'project' above. Imagine that we are sitting on one atom in the unit cell of our crystal and that there is just one atomic orbital per atom. (This is atom *n*.) Then, we want to scan over *all* energies of the system to determine whether $\mathscr{E} = \mathscr{E}_k$. If this condition is satisfied, then we determine $\left(\frac{dS}{d\mathscr{E}} \right)_{\mathscr{E}=\mathscr{E}_k}$ and weight it by the appropriate factor. To recover the total DOS, we just add up the contribution over all basis states *j*. For the local DOS, $\left(\frac{dS}{d\mathscr{E}} \right)_{\mathscr{E}=\mathscr{E}_k}$ has a slightly subtle interpretation. Namely, it only gets 'switched on' when we find an energy

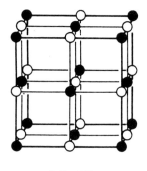

NaCl

Fig. 9.9. The structure of a crystal of NaCl showing the Na^+ ions (black spheres) and the Cl^- ions (white spheres).

for which $\mathscr{E} = \mathscr{E}_k$. Therefore,

$$\left(\frac{dS}{d\mathscr{E}}\right)_{\mathscr{E}=\mathscr{E}_k} = \delta(\mathscr{E} - \mathscr{E}_k) \tag{9.20}$$

is a Dirac delta function. Properties of the Dirac delta function are reviewed in Appendix 5.

The local DOS is particularly useful when we are trying to determine the contribution of a given type of atom to the total DOS of a crystal. For example, consider the simple NaCl crystal, the structure of which is shown in Fig. 9.9. Recall from Ch. 1 that Cl has a higher electronegativity than Na. Therefore, in the total DOS of NaCl, we would find by examining the local DOS that: (1) the bottom of the DOS is centered on the Cl atom and (2) the top of the DOS is centered on the Na atom. These findings reflect the fact that there is a close *link* between the local atomic environment and the local DOS.

9.4 The band energy and the bond energy

Before we explore the link between the local DOS and the atomic environment, let us make a brief diversion to investigate another important aspect of the density matrix. The *band energy*, $\mathscr{E}_{band}$, is the sum of the occupied eigenvalues of the system,

$$\mathscr{E}_{band} = 2 \sum_{n \, occ} \mathscr{E}_n \tag{9.21}$$

where, of course, the factor of 2 arises from spin up and spin down electrons in each band. For a system with a large number of atoms, we can replace the sum by an integral,

$$\mathscr{E}_{\text{band}} = 2 \int^{\mathscr{E}_F} \mathscr{E}\, D(\mathscr{E})\, d\mathscr{E} \qquad (9.22)$$

The DOS appears because it is the distribution of electrons in energy. The limits of the integral are from the bottom of the band to $\mathscr{E}_F$. We are still working under the assumption that the temperature is 0 K. The band energy is a measure of the cohesion of the solid, or how much the system's energy is lowered if the atoms are brought together from infinite separation.

We mentioned in the previous section that there was some question about the zero of energy in the DOS. The on-site Hamiltonian matrix element, α, is actually the energy against which all bonding is measured. Recall that every expression we found for a band energy always had as the first term α, and additional terms related to the hopping matrix elements. To discuss the energetics of bonding, then, it makes more sense to consider a *bond* energy,

$$\mathscr{E}_{\text{bond}} = 2 \int^{\mathscr{E}_F} (\mathscr{E} - \alpha)\, D(\mathscr{E})\, d\mathscr{E} \qquad (9.23)$$

which may be rewritten as

$$\begin{aligned}
\mathscr{E}_{\text{bond}} &= 2 \int^{\mathscr{E}_F} \mathscr{E}\, D(\mathscr{E})\, d\mathscr{E} - 2\alpha \int^{\mathscr{E}_F} D(\mathscr{E})\, d\mathscr{E} \\
&= \mathscr{E}_{\text{band}} - N_e \alpha \qquad (9.24)
\end{aligned}$$

where $\mathscr{E}_{\text{band}}$ is the band energy from Eq. 9.22 and N_e is the number of electrons in the system.

Let us now develop Eq. 9.24 more rigorously, calling on our discussion of bond energy and bond order in molecules at the end of Ch. 7. For convenience, we will do this in the context of a molecular system that has a very large number of atoms. Thus,

$$\mathscr{E}_{\text{bond}} = 2 \sum_{k \text{ occ}} (\mathscr{E}_k - \alpha) \qquad (9.25)$$

where $\mathscr{E}_k$ is a solution to the Schrödinger equation,

$$\begin{aligned}
\mathscr{E}_k &= \langle \psi | \mathscr{H} | \psi \rangle \\
&= \sum_i \sum_j \langle \psi | i \rangle \langle i | \mathscr{H} | j \rangle \langle j | \psi \rangle \qquad (9.26)
\end{aligned}$$

The second expression results from inserting the closure relation (Eq. 4.14) between two complete sets of states $\{|i\rangle\}$ and $\{|j\rangle\}$. Substituting Eq. 9.26 into Eq. 9.25 yields

$$
\begin{aligned}
\mathscr{E}_{\text{bond}} &= 2\sum_i \sum_j \sum_{k\,\text{occ}} \langle\psi|i\rangle(\langle i|\mathscr{H}|j\rangle - \alpha\langle i|j\rangle)\langle j|\psi\rangle \\
&= \sum_i \sum_j \left(2\sum_{k\,\text{occ}} \langle j|\psi\rangle\langle\psi|i\rangle\right)(\langle i|\mathscr{H}|j\rangle - \alpha\langle i|j\rangle) \qquad (9.27)
\end{aligned}
$$

Note that the term in parentheses is just the density matrix element, ρ_{ji}, Eq. 9.10. Recognizing that $\langle i|j\rangle = \delta_{ij}$ and $\langle i|\mathscr{H}|i\rangle \equiv \mathscr{H}_{ii} = \alpha$, we can rewrite Eq. 9.27 as

$$
\mathscr{E}_{\text{bond}} = \sum_i \sum_{j\neq i} \rho_{ji}\mathscr{H}_{ij} \qquad (9.28)
$$

Furthermore, it can be shown that

$$
\mathscr{E}_{\text{band}} = \sum_i \sum_j \rho_{ji}\mathscr{H}_{ij} \qquad (9.29)
$$

For a crystal with a very large number of atoms, the sums are replaced by integrals.

Exercise 9.5   Derive Eq. 9.29 from Eq. 9.21.

These two expressions are rather innocuous-looking, but they are in fact extremely useful. We can consider Eqs. 9.28 and 9.29 as transformations between reciprocal space, wherein eigenstates dominate, and real space, wherein charge densities and atomic orbitals dominate. Most notably, these two equations do not rely on translational symmetry, which means that we can use them to characterize and understand systems lacking long-range order. The key to implementing this approach, however, lies in determining the density matrix elements. More specifically, we would like to be able to do this without having recourse to determining the eigenstates of the system. A beginning in this direction is sketched in the next section.

9.5 The moments theorem

Earlier in this chapter we introduced the DOS, which is a distribution function for the energy of the electrons in a system. One main advantage of working with the DOS is that it allows us to examine bonding in systems for which the concept of reciprocal space breaks down or for systems in which calculation of the band structure is difficult. An additional advantage of working with a distribution function is that, in many cases, we do not need an exact expression for the function itself. Many properties of the distribution are characterized simply by knowing the *moments* of the distribution. For more about mathematical details of moments, see Appendix 6.

We will now introduce a theorem that was first derived by Cyrot-Lackmann in 1968.† This 'moments theorem' relates the local DOS to the topology of the local environment. Consider the local DOS for an atom labeled j. From the expression for the local DOS, Eq. 9.19, we have

$$d_j(\mathscr{E}) = \sum_{\text{all eigenvalues } k} \left(\frac{\mathrm{d}S}{\mathrm{d}\mathscr{E}}\right)_{\mathscr{E}=\mathscr{E}_k} |\langle j|\psi\rangle|^2$$

$$= \sum_{\text{all eigenvalues } k} \left(\frac{\mathrm{d}S}{\mathrm{d}\mathscr{E}}\right)_{\mathscr{E}=\mathscr{E}_k} \langle j|\psi\rangle\langle\psi|j\rangle \qquad (9.30)$$

Now, the nth moment of the local DOS about the on-site matrix element $\mathscr{H}_{jj} \equiv \alpha$ is given by:

$$\mu_j^{(n)} = \int_{\text{whole band}} (\mathscr{E} - \mathscr{H}_{jj})^n \, d_j(\mathscr{E}) \, \mathrm{d}\mathscr{E} \qquad (9.31)$$

The zeroth moment is 1 because it is simply the integral of the local DOS over the entire band. The first moment is the center of gravity of the local DOS, or the average, relative to $\mathscr{H}_{jj}$. The second moment, $\mu_j^{(2)}$, is the variance of the local DOS. The square root of $\mu_j^{(2)}$ is a measure of the width of the local DOS. The third moment measures the skewness of the local DOS about the center of gravity. A large negative value of $\mu_j^{(3)}$ corresponds to a long tail in the local DOS below the center of gravity accompanied by a compressed peak above the center of gravity. The fourth moment measures the tendency for a gap to form in the middle of the band. A negative value of the function $s \equiv \mu_j^{(4)}/(\mu_j^{(2)})^2 - 1$ is consistent with two well-localized peaks, while a positive value corresponds to one central peak. As we can see, then, the higher moments give a great deal of information about the shape of the local DOS.

Inserting Eq. 9.30 into Eq. 9.31 yields

$$\mu_j^{(n)} = \int_{\text{whole band}} \left(\sum_{\text{all eigenvalues } k} (\mathscr{E} - \mathscr{H}_{jj})^n \left(\frac{\mathrm{d}S}{\mathrm{d}\mathscr{E}}\right)_{\mathscr{E}=\mathscr{E}_k} \langle j|\psi\rangle\langle\psi|j\rangle \mathrm{d}\mathscr{E} \right) \qquad (9.32)$$

for the nth moment of the local DOS. Recall from Eq. 9.20 that the contribution to the total DOS for the kth eigenstate is just a Dirac delta function.

† F. Cyrot-Lackmann, *J. Phys. Chem. Solids* **29**, 1235 (1968).

Hence, all terms in the integration are zero except when $\mathcal{E} = \mathcal{E}_k$, so that

$$\mu_j^{(n)} = \sum_{\text{all eigenstates}} \langle j|\psi\rangle(\mathcal{E}_k - \mathcal{H}_{jj})^n\langle\psi|j\rangle \tag{9.33}$$

where the sum is now over all eigenstates. We can extract the identity operation for the complete set of states $\{|\psi\rangle\}$ to obtain the more compact expression for the moments,

$$\mu_j^{(n)} = \langle j|(\mathcal{H} - \mathcal{H}_{jj})^n|j\rangle \tag{9.34}$$

Using this expression, we can readily calculate some of the lower moments. For example, $\mu_j^{(0)} = \langle j|j\rangle = 1$, while $\mu_j^{(1)} = \langle j|(\mathcal{H} - \mathcal{H}_{jj})|j\rangle = \mathcal{H}_{jj} - \mathcal{H}_{jj}\langle j|j\rangle = 0$. The center of gravity of a band is thus always $\mathcal{H}_{jj}$. To evaluate the second moment requires a little subtle manipulation. Write out the expression for the second moment,

$$\begin{aligned}\mu_j^{(2)} &= \langle j|(\mathcal{H} - \mathcal{H}_{jj})^2|j\rangle \\ &= \langle j|(\mathcal{H} - \mathcal{H}_{jj})(\mathcal{H} - \mathcal{H}_{jj})|j\rangle\end{aligned} \tag{9.35}$$

and insert a complete set of basis states $\{|i\rangle\}$,

$$\mu_j^{(2)} = \sum_i \langle j|\mathcal{H} - \mathcal{H}_{jj}|i\rangle\langle i|\mathcal{H} - \mathcal{H}_{jj}|j\rangle \tag{9.36}$$

If we assume that only nearest-neighbor interactions are nonzero, then only those atoms i that are neighbors of atom j have nonzero Hamiltonian matrix elements $\mathcal{H}_{ji}$. Furthermore, $\langle j|\mathcal{H} - \mathcal{H}_{jj}|j\rangle = 0$, and $\langle j|\mathcal{H} - \mathcal{H}_{jj}|i\rangle = \mathcal{H}_{ji}$ for $j \neq i$. Thus, we can simplify the expression for the second moment to read

$$\mu_j^{(2)} = \sum_{i \neq j} \mathcal{H}_{ji}\mathcal{H}_{ij} \tag{9.37}$$

Each term $\mathcal{H}_{ji}\mathcal{H}_{ij}$ describes the energy of an electron starting at site j, hopping out to a neighboring site i, and hopping back to j. This is a path of length 2 hops from atom j. The second moment of the local DOS is thus the sum of all such paths of length 2 hops. In a perfect crystal, where the coordination number is z and the nearest neighbor hopping integral is β, the second moment of the local DOS is $z\beta^2$. The root mean square width of the band is proportional to $\sqrt{z}$.

It is a straightforward task to generalize this result to higher moments. For example, the third moment is given by

$$\mu_j^{(3)} = \sum_{i \neq j}\sum_{l \neq j} \mathcal{H}_{ji}\mathcal{H}_{il}\mathcal{H}_{lj} \tag{9.38}$$

which is the sum of all paths of length 3 hops starting and ending on site j. The fourth moment is given by

$$\mu_j^{(4)} = \sum_{i \neq j} \sum_{l \neq j} \sum_{l \neq k} \mathcal{H}_{ji} \mathcal{H}_{ij} \mathcal{H}_{lj} \mathcal{H}_{jk} \qquad (9.39)$$

Note that in determining moments higher than the second, paths that consist of hops back and forth between two atoms are allowed.

Exercise 9.6   Show the last step to arrive at Eq. 9.38.

Exercise 9.7   Apply the moments theorem to the simple one-dimensional ring of H atoms to find the moments up to the third moment of the DOS distribution function. Compare your findings with the results in Fig. 9.7 and Eq. 9.16.

We are now in a position to state the moments theorem:

Theorem 9.1 *Within the nearest-neighbor interaction approximation, the nth moment of the local density of states on atom j is the sum of all paths of length n hops starting and ending at site j.*

This is a very remarkable result. Briefly, let us think about what it means. As we have seen above, even for the simple ring of H atoms, it is a somewhat tedious process to calculate the band structure and from that to derive the local density of states. But, using the moments theorem, we can derive exact expressions for the *moments* of the local density of states which tell us about the shape, skewness, *etc.*, of the distribution function. In turn, this information about the distribution function can lead us to insightful information about the electronic structure of various atoms throughout our system. For example, we can determine whether states on one atom are more or less localized than those on another atom in a different part of the solid. We can examine the local density of states on atoms located, for instance, at a grain boundary and compare it with a bulk atom's density of states. Most importantly, the specific characteristics of the distribution function are derived from examination of the *real space* environment of the atom of interest. Thus, we have formed a direct link between real space and reciprocal space. We will come back to use the moments theorem quite a lot in the remainder of the book.

Problems

9.1   Return to the infinite linear chain of A and B atoms with periodic boundary conditions, Problem 8.4. Using the band structure that you

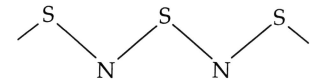

Fig. 9.10. Zig-zag chain of S and N atoms with 120° bond angles.

found in that problem, sketch the total DOS and the local DOS, $d_A(\mathscr{E})$ and $d_B(\mathscr{E})$. For a specified Fermi energy do you expect the amount of electronic charge on an A atom to be the same as that on a B atom? If not, how can you calculate the difference? Sketch the bond order between neighboring atoms as $\mathscr{E}_F$ varies from the bottom of the bands to the top of the bands.

9.2   Reconsider the infinite ring of H atoms. Break the ring so that atom 1 is not connected to atom N. Use the moments theorem to find the first three moments of atom 1. Compare your results with the infinite ring.

9.3   The first 'inorganic' polymer discovered to be a metallic conductor in its pure state was polysulfurnitride, $(SN)_x$. It forms a one-dimensional zig-zag chain with bond angles of 120°, as shown in Fig. 9.10.

The primary bonding in this material comes from σ bonds between the neighboring S and N atoms. In addition, there are nonbonding levels to which each atom donates two electrons. The remaining three total (one from nitrogen and two from sulfur) valence electrons are free to participate in π bonding above and below the plane of the zig-zag chain. The π bonds are formed from p_z orbitals which are perpendicular to the page.

  (a) Considering only these p_z orbitals, what is the band structure of $(SN)_x$? The first ionization potentials of S and N are 14.53 eV and 8.152 eV, respectively.

  (b) Draw the DOS for these bands. How do the local DOS for N and S differ?

  (c) Apply the moments theorem to this solid and compare the results for the first three moments of the local DOS with your results above.

9.4   What can you say about the moments of the DOS for one-dimensional systems? How might you obtain a DOS that is skewed?

9.5   Exercise 9.1.

9.6   Exercise 9.3.

9.7   Exercise 9.4.

9.8   Exercise 9.6.

9.9   Consider a system with translational symmetry that may be one-, two-, or three-dimensional. Assume that both the on-site Hamiltonian matrix elements and the hopping integrals are independent of small displacements. Derive an expression for the derivative of the *band* energy with respect to the x coordinate, for example, of an atom m. Your result is known as the Hellman–Feynman theorem and it is used widely to calculate the forces exerted by one atom on its neighbors.

10

Generalization to two and three dimensions

10.1 Introduction

One-dimensional systems serve as important pedagogical tools for developing concepts and intuition about electronic structure. However, there are also important examples of real one-dimensional systems. We can think of individual polymer chains, ultra-small 'wires' of near-atomic widths made by advanced lithography techniques, as well as quasi-one-dimensional systems like polyacetylene, which undergoes a Peierls distortion, a topic to which we will return later. Nonetheless, there is a broader horizon in two and three dimensions.

Indeed, there is a great deal of attention focused on surfaces as two-dimensional structures. Surfaces are interesting in their own right because of their structure and any possible reconstruction that may occur. One challenging theoretical problem that has been recently solved is the famous 7×7 reconstruction of the Si(111) surface.† With the advent of new microscopies such as Scanning Tunneling Microscopy (STM) and Atomic Force Microscopy (AFM), we might be interested in how the atoms in the probe tip interact with the surface. A number of chemical reactions are also mediated by surfaces, catalysis being one of the most common examples. Finally, though, some three-dimensional systems may actually be more two-dimensional in nature. For example, graphite consists of sheets with strong in-plane bonding but weak inter-plane bonding. The electronic structure of the high-T_c superconductors is also dominated by the presence of Cu–O planes.

Of course, there is no denying the preponderance of three-dimensional systems, ranging from electronic materials like Si and GaAs to metals like

---

† See, for example, G.-X. Qian and D.J. Chadi, *Phys. Rev. B* **35**, 1288 (1987).

elemental metals, metallic alloys, or superalloys to ionic, insulating ceramics like Al_2O_3 and other oxides to amorphous materials. The list is long.

The foundations for many concepts in electronic structure in previous chapters have been laid in one dimension. In this chapter, we generalize these concepts to higher dimensions. In some ways, the difficulties with this generalization lie less with the mathematics than with the complexities of visualization and interpretation. In this case, the real utility of the local DOS and the moments theorem will become apparent.

10.2 The Schrödinger equation

We introduced the Schrödinger equation in Ch. 5 and discussed its solution in detail for an isolated H atom in Ch. 6. There we saw that the three-dimensional problem of finding the eigenvalues and eigenvectors could be broken down into solving three one-dimensional problems using separation of variables. All of the molecular systems and simple crystals that we have solved the Schrödinger equation for have been one-dimensional.

Consider the Schrödinger equation for a general two-dimensional system,

$$\mathscr{H}\psi = \mathscr{E}\psi$$

$$-\frac{\hbar^2}{2m}\left(\frac{\partial^2}{\partial x^2} + \frac{\partial^2}{\partial y^2}\right)\psi(x,y) + V(x,y)\psi(x,y) = \mathscr{E}\psi(x,y) \quad (10.1)$$

Letting

$$\psi(x,y) = X(x)Y(y) \quad (10.2)$$

we find

$$-\frac{\hbar^2}{2m}\left(\frac{\partial^2}{\partial x^2} + \frac{\partial^2}{\partial y^2}\right)X(x)Y(y) + V(x,y)X(x)Y(y) = \mathscr{E}X(x)Y(y) \quad (10.3)$$

If the potential is a function only of the interatomic distance, then Eq. 10.3 is separable in the two spatial dimensions and we obtain two identical equations for the x and y dependence,

$$-\frac{\hbar^2}{2m}\frac{d^2}{dx^2}X(x) + V(|\vec{r}|)X(x) = \mathscr{E}_x X(x)$$

$$-\frac{\hbar^2}{2m}\frac{d^2}{dy^2}Y(y) + V(|\vec{r}|)Y(y) = \mathscr{E}_y Y(y) \quad (10.4)$$

and $\mathscr{E}_x + \mathscr{E}_y = \mathscr{E}$.

Precisely the same arguments can be applied in three dimensions. In that

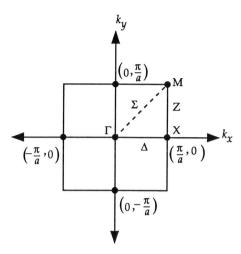

Fig. 10.1. The first Brillouin zone of a square lattice in two dimensions. Special points and lines are identified in the figure.

case, we obtain three separable equations, two identical with Eqs. 10.4 plus an equation for the z direction.

In Ch. 8, we discussed the translation group of a perfect crystal with periodic boundary conditions. There we showed that Bloch's theorem was a consequence of translational symmetry and that the Bloch wavefunctions are just the functions that generate the irreducible representations of the translation group. We found that each irreducible representation was labeled by a quantity k, the wavevector. In one dimension, the bounds on allowed k values were easy to find. In two and three dimensions, however, we must explicitly recognize that $\vec{k}$ is indeed a vector. Because there is a greater variety of unit cells possible, the reciprocal lattice vectors and the first Brillouin zones will, in general, have more complicated forms.

To illustrate these ideas, let us consider a two-dimensional square lattice. Let the spacing between lattice points be a in each direction and assume that each point is occupied by a single s atomic orbital. Since the two spatial dimensions are separable and identical, the two components of the wavevector are restricted to $-\pi/a \le k_x \le \pi/a$ and $-\pi/a \le k_y \le \pi/a$. Unlike in one dimension, there are now N^2 values of k confined to this first Brillouin zone, which is shown in Fig. 10.1.

Let us examine the symmetry of various points throughout the Brillouin zone. The point at the center of the zone is always given the label Γ and it is called, appropriately enough, the 'Γ point'. It possesses the full symmetry of a square: $\{E, 2C_4, C_2, 2\sigma_v, 2\sigma_d\}$, where the rotation axis passes through the Γ point and is perpendicular to the plane of the square. There is no σ_h operation because this is a two-dimensional object. The point at the corner of the Brillouin zone is called M and, for the same origin, it has the same set of symmetry operations as the Γ point. The point in the middle of one of the faces is called X and it has lower symmetry than either Γ or M. The symmetry operations of this point are $\{E, C_2, \sigma_v^x, \sigma_v^y\}$. The fact that the X point has lower symmetry is not immediately apparent. At first sight, it appears as if X is at a position of four-fold symmetry as well. However, under a C_4 rotation, the point $(\pi/a, 0)$ transforms into the point $(0, \pi/a)$. But these two points are not related by a reciprocal lattice vector,† so they *cannot be equivalent*. Reflection across a mirror plane coincident with the y axis gives $(\pi/a, 0) \rightarrow (-\pi/a, 0)$, which are related by the reciprocal lattice vector $\vec{k} = 2\pi/a\hat{x}$. Any general point in the Brillouin zone possesses only $\{E\}$.

Notice that *lines* in the Brillouin zone can also have symmetry; in other words, all points along the line have the same set of symmetry operations. There are three unique, nontrivial lines in the square Brillouin zone and they are:

$$\Delta \quad \{E, \sigma_v^y\}$$
$$\Sigma \quad \{E, \sigma_d\}$$
$$Z \quad \{E, \sigma_v^x\}$$

These symbols are part of the Bouckaert–Smoluchowski–Wigner (BSW) notation, which is often used in solid state physics for labeling points and, hence, energy bands in the Brillouin zone.

Exercise 10.1   Verify that an arbitrary point along any of the lines Δ, Σ, or Z does possess the symmetry operations so indicated.

What are the implications of the fact that we have special points and lines? At these points and along these lines, there are some degeneracies that are dictated *solely* by symmetry and are not lifted no matter how strong the potential energy. Furthermore, the different groups along special lines or at special points will tell us about how the various wavefunctions transform. There are three important points to keep in mind:

(i) Wavefunctions that transform as different irreducible representations

---

† From Ch. 8, reciprocal lattice vectors for the two-dimensional square lattice are of the form $\vec{k} = n_1\hat{b}_1 + n_2\hat{b}_2$, where $\hat{b}_1 = (2\pi/a)\hat{x}$ and $\hat{b}_2 = (2\pi/a)\hat{y}$.

have different energies, except that their energy levels may cross. If these bands do cross, it is called an *accidental degeneracy*.

(ii) Wavefunctions that transform as the same irreducible representation repel one another and they may not cross.

(iii) Some irreducible representations are of dimension two or three, so that these bands are degenerate.

Let us turn to the calculation of the wavefunctions and the band structure for the simple square lattice. As we did in one dimension, our wavefunction can be written within the tight-binding approximation,

$$|\psi\rangle = \sum_m \sum_n c_{mn} |\phi_{mn}\rangle \tag{10.5}$$

where the integers m and n index positions in the square lattice in the x and y directions, respectively. From Bloch's theorem, we know the coefficients are of the form

$$c_{mn} = \exp(\imath mk_x a + \imath nk_y a) \tag{10.6}$$

so that a normalized wavefunction is

$$|\psi\rangle = \frac{1}{\sqrt{N}} \sum_m \sum_n e^{\imath mk_x a} e^{\imath nk_y a} |\phi_{mn}\rangle \tag{10.7}$$

with N the total number of sites in the lattice. Schematic pictures of the real parts of the wavefunctions for specific k points are drawn in Fig. 10.2. Recall that it is perfectly acceptable for a wavefunction to have an imaginary part, since what we observe is $||\psi\rangle|^2$, which will always be real.

Inserting the wavefunction of Eq. 10.7 into the Schrödinger equation and trying to set up and solve the secular determinant as we did in Ch. 8 for the one-dimensional case is actually quite cumbersome. Now, Hamiltonian matrix elements are labeled with two pairs of indices. It is more transparent to find the band structure using a slightly more direct technique. Instead, let us write

$$\mathcal{H}|\psi\rangle = \frac{1}{\sqrt{N}} \sum_m \sum_n e^{\imath mk_x a} e^{\imath nk_y a} \mathcal{H} |\phi_{mn}\rangle \tag{10.8}$$

$$\mathcal{E}|\psi\rangle = \frac{\mathcal{E}}{\sqrt{N}} \sum_m \sum_n e^{\imath mk_x a} e^{\imath nk_y a} |\phi_{mn}\rangle \tag{10.9}$$

Left-multiply each equation by some wavefunction

$$|\psi'\rangle = \frac{1}{\sqrt{N}} \sum_p \sum_q e^{\imath pk_x a} e^{\imath qk_y a} |\phi_{pq}\rangle \tag{10.10}$$

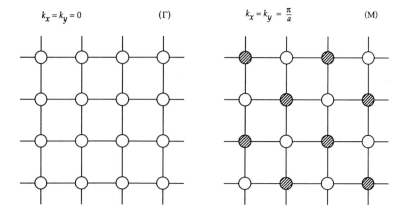

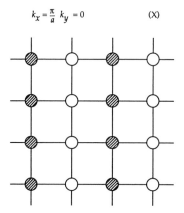

Fig. 10.2. Real-space sketches of the wavefunctions for the two-dimensional one s orbital per atom square lattice. Only the real parts of the wavefunctions are shown at the k points indicated. Shaded atoms have the opposite phase to unshaded atoms.

such that

$$\langle\psi'|\mathcal{H}|\psi\rangle = \frac{1}{N}\sum_{mn}\sum_{pq} e^{i(m-p)k_xa}e^{i(n-q)k_ya}\langle\phi_{pq}|\mathcal{H}|\phi_{mn}\rangle \qquad (10.11)$$

$$\mathcal{E}\langle\psi'|\psi\rangle = \frac{\mathcal{E}}{N}\sum_{mn}\sum_{pq} e^{i(m-p)k_xa}e^{i(n-q)k_ya}\langle\phi_{pq}|\phi_{mn}\rangle \qquad (10.12)$$

Invoking the neglect of overlap approximation means that

$$\langle\phi_{pq}|\phi_{mn}\rangle = \delta_{pm}\delta_{qn} \qquad (10.13)$$

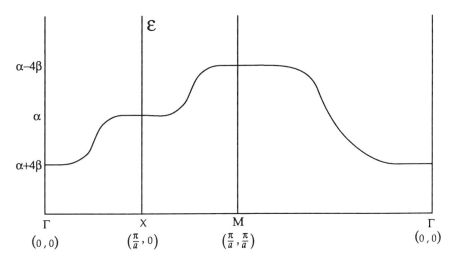

Fig. 10.3. Calculated band structure of the one s orbital per atom square lattice plotted around a circuit of special lines.

Making the Hückel approximation so that on-site Hamiltonian matrix elements are denoted by α and adjacent atoms' Hamiltonian matrix elements are β, with *all* other matrix elements equal to zero, we find that there are only five nonzero terms in the summation of Eq. 10.11; one on-site term and four adjacent-neighbor terms,

$$\langle \psi' | \mathscr{H} | \psi \rangle = \alpha + [e^{ik_x a}\beta + e^{ik_y a}\beta + e^{-ik_x a}\beta + e^{-ik_y a}\beta]$$
$$= \alpha + 2\beta[\cos(k_x a) + \cos(k_y a)] \qquad (10.14)$$

Exercise 10.2   Show that Eq. 10.12 reduces to $\langle \psi' | \psi \rangle = 1$.

Therefore,

$$\mathscr{E} = \alpha + 2\beta[\cos(k_x a) + \cos(k_y a)] \qquad (10.15)$$

The energy has a form similar to the one-dimensional case except that now we are working in an infinite plane.

It is clearly impossible to plot this band structure at every point $\vec{k}$ in the first Brillouin zone. Fortunately, however, we can rely on symmetry to simplify our task. What is most often done is to plot the band structure around a closed path of special lines. In our case, it is instructive to use the circuit $\Gamma \xrightarrow{\Delta} X \xrightarrow{Z} M \xrightarrow{\Sigma} \Gamma$ from Fig. 10.1. This circuit gives a representative sampling of the band structure throughout the entire first zone. It is plotted in Fig. 10.3.

Compare the eigenvalues displayed in Fig. 10.3 with the corresponding wavefunctions in Fig. 10.2. The lowest-energy point in the Brillouin zone is at the Γ point and this corresponds to the fully bonding wavefunction. Likewise, the M point is highest in energy and it corresponds to the fully antibonding wavefunction. The X point is in between the two extremes, at $\mathscr{E} = \alpha$, actually. Note that the corresponding wavefunction is bonding in the k_y direction and antibonding in the k_x direction. Overall, the wavefunction is *nonbonding*, which is reflected in the fact that the energy at this point is just the on-site Hamiltonian matrix element.

Just as in one dimension, we have the useful concept of the Fermi energy, $\mathscr{E}_F$, which separates filled from unfilled states, and the accompanying Fermi wave vector, k_F. However, instead of being just a straight line, as in one dimension, the Fermi energy becomes a closed curve or 'surface' in higher dimensions. For the square lattice occupied by s-type orbitals, the Fermi energy is defined by

$$\mathscr{E}_F = \alpha + 2\beta[\cos(k_{F,x}a) + \cos(k_{F,y}a)] \tag{10.16}$$

When the band is nearly empty, $k_{F,x}$ and $k_{F,y}$ are small and the cosines in Eq. 10.16 can be expanded,

$$\mathscr{E}_F = \alpha + 4\beta - \beta a^2(k_{F,x}^2 + k_{F,y}^2) \tag{10.17}$$

which is just the equation for a circle in reciprocal space. As we add more electrons to the band, the circle bulges out until at a half-filled band, it finally intersects the Brillouin zone boundaries at four points $(\pi/a, 0), (0, \pi/a)$, $(-\pi/a, 0)$, and $(0, -\pi/a)$. For a nearly full band, only the regions near the corners at the four equivalent M points remain unoccupied. Schematic drawings for these three band fillings are shown in Fig. 10.4. The occupied regions are shown shaded.

10.3 The density of states

We introduced the concept of the density of states in Ch. 9 both because it codified a lot of information from the complicated band structure into a single distribution function and because this distribution function was seen to be the essential link between real space and reciprocal space through the moments theorem. We can sketch the DOS for the square lattice just as we did for the one-dimensional ring. In the present case, however, inspection of the band structure of Fig. 10.3 indicates that the energy for which the DOS diverges is at the X point. At Γ and M, the DOS approaches a finite value. The singularity arises because the slope of $\mathscr{E}(\vec{k})$ at X vanishes.

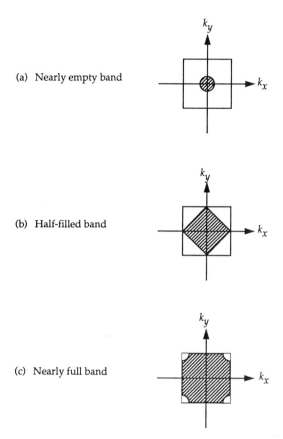

(a) Nearly empty band

(b) Half-filled band

(c) Nearly full band

Fig. 10.4. Schematic diagram showing the filled (shaded) and unfilled bands in the first Brillouin zone of one atom per site on a two-dimensional square lattice.

How do we show these results analytically? Technically, all we need do is recognize that the wavevector is now truly a vector. Following the discussion in Ch. 9, it is still valid to write that

$$D(\mathscr{E})d\mathscr{E} = \text{number of states per unit volume between } \mathscr{E} \text{ and } \mathscr{E}+d\mathscr{E} \quad (10.18)$$

or, from Eq. 9.15,

$$D(\mathscr{E}) = 2 \left(\frac{\text{volume of } k\text{-space unit cell where } \mathscr{E} \leq \mathscr{E}(\vec{k}) \leq \mathscr{E} + d\mathscr{E}}{\text{volume per allowed wavevector}} \right) \quad (10.19)$$

The (two-dimensional) volume of a k-space unit cell with the constraint $\mathscr{E} \leq \mathscr{E}(\vec{k}) \leq \mathscr{E} + d\mathscr{E}$ is $\int d\vec{k}\, \delta(\mathscr{E} - \mathscr{E}(\vec{k}))$. The volume per allowed wavevector is $(2\pi/Na)^2$. The volume integral over the delta function can be rewritten as

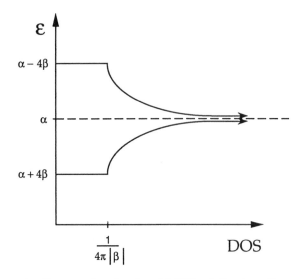

Fig. 10.5. Schematic diagram for the local DOS of the two-dimensional square lattice occupied by one *s* atomic orbital per site.

a surface integral,

$$D(\mathscr{E}) = 2\left(\frac{Na}{2\pi}\right)^2 \int_{S(\mathscr{E})} dS\, \frac{1}{|\nabla_k \mathscr{E}(\vec{k})|} \tag{10.20}$$

which can be recast as a density of states per lattice site (local DOS) by dividing through by N^2,

$$
\begin{aligned}
d(\mathscr{E}) &= \frac{D(\mathscr{E})}{N^2} \\
&= \frac{a^2}{2\pi^2} \int_{S(\mathscr{E})} dS\, \frac{1}{|\nabla_k \mathscr{E}(\vec{k})|}
\end{aligned}
\tag{10.21}
$$

With the band structure given by Eq. 10.15, it is difficult to find a general analytic expression for the local DOS. However, for several limiting cases, the integral is tractable.

Exercise 10.3 For a nearly empty or a nearly full band, show that the local DOS is $d(\mathscr{E}) = 1/4\pi|\beta|$.

Exercise 10.4 For a half-full band, show that the local DOS is infinite.

Thus, we arrive at the schematic diagram in Fig. 10.5 for the local DOS of the two-dimensional square lattice occupied by one *s* atomic orbital per site.

We can use the same argument as above to find the local DOS for a three-dimensional simple cubic crystal. In that case, we do not find any singularities. Still, from the functional form of Eq. 10.21, there are discontinuities, called *van Hove singularities*, in the local DOS whenever the slope of the band structure changes discontinuously.

10.4 The moments theorem in two and three dimensions

The moments theorem, as introduced in Ch. 9, needs no rewriting or reinterpretation when we move on to higher dimensions. Thus, it still provides the most direct link between the local coordination and topology of an atom in a solid and information about the local electronic structure. The only difficulty now, of course, is that the number of paths of different lengths increases rapidly and there is no *bona fide* way of guaranteeing that our counting is complete. We must therefore remember to use the moments theorem systematically.

Let us demonstrate its use in the two-dimensional square lattice. As for the one-dimensional case, $\mu_j^{(0)} = 1$ and $\mu_j^{(1)} = 0$, so that the local DOS is centered about $\mathscr{E} = \alpha$. Furthermore, $\mu_j^{(2)} = 4\beta^2$, since there are now four paths of length two hops (Fig. 10.6a). This means that the band width for the two-dimensional square lattice is larger than for the chain. All odd moments will be zero, so that the band is symmetrical about $\mathscr{E} = \alpha$. Finally, the fourth moment (Fig. 10.6b) is given by $\mu_j^{(4)} = 24\beta^4$, so that the function s is $\mu_j^{(4)}/(\mu_j^{(2)})^2 - 1 = \frac{1}{2}$. Therefore, the DOS is unimodal. Compare the features of the local DOS derived using the moments theorem with the plot in Fig. 10.5.

10.5 The density matrix, band energy, and bond energy in higher dimensions

In Section 9.2, we derived an expression for the density matrix of a one-dimensional system. Formally, the derivation follows identically for two and three dimensions, with only the density of states in k space changing. We also relabel the atom positions with vectors $\vec{R}_m$ for atom m, for example. Therefore, for a three-dimensional system, the density matrix elements are given by

$$\rho_{mn} \equiv 2 \left(\frac{a}{2\pi} \right)^3 \int_{-k_F}^{k_F} d\vec{k} \, e^{i\vec{k} \cdot (\vec{R}_m - \vec{R}_n)} \tag{10.22}$$

Exercise 10.5   Using the fact that $d\vec{k} = k^2 dk \sin\theta d\theta d\phi$, find a general expression for Eq. 10.22. Use your resulting expression to

(a) Paths of 2 hops length = 4

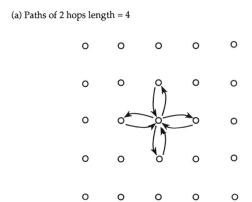

(b) Paths of 4 hops length = 24

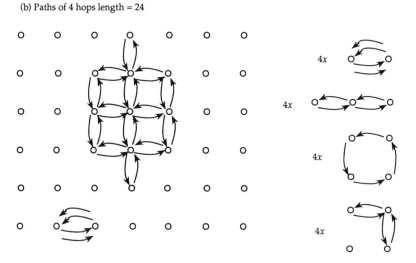

Fig. 10.6. Schematic diagram of a two-dimensional lattice used for calculating the number of closed paths of different lengths. (a) Paths of length 2; there are four such paths. (b) Paths of length 4; there are 24 such paths.

evaluate the diagonal and off-diagonal elements of the density matrix and hence find, respectively, the charge per atom and the bond order.

In Section 9.4, we developed expressions relating the band energy (the sum over the energy eigenvalues $\mathscr{E}_{band}$), the bond energy (the band energy measured relative to the diagonal Hamiltonian matrix elements $\mathscr{E}_{bond}$), and the density matrix elements. Since Eqs. 9.28 and 9.29 do not rely on dimensionality, they carry over identically to higher dimensions.

Problems

10.1 Consider an infinite two-dimensional square lattice of H atoms. Remove one H atom to create a vacancy. Use the moments theorem to find the first three moments of the local DOS for one of the atoms bordering the vacancy. Compare your result with the infinite two-dimensional perfect square lattice. What can you conclude about the local DOS for an atom that is on a 'surface'?

10.2   (a) Place p-type atomic orbitals on all lattice sites of a two-dimensional square lattice. Assume that they do not overlap with any s-type atomic orbitals. Write down an expression for the wavefunction and find the band structure at the special points throughout the first Brillouin zone.

    (b) Now stretch the lattice in one direction so that we obtain a rectangular lattice with $a_1 = 2a_2$. Find the special points in the Brillouin zone and identify their symmetry operations. Write down an expression for the total wavefunction of this system, assuming that each site is occupied by p-type orbitals. Sketch the band structure of this system and compare it with that in the first part of this question. What happens in the limit $a_1 \gg a_2$?

10.3 Consider an infinite square lattice in two dimensions with d orbitals on each site. There are no interactions with s or p orbitals. Find the band structure of this system and sketch it schematically.

10.4 Apply the moments theorem to an infinite *hexagonal* network of atoms in two dimensions. That is, each atom has six identical neighbors. Find the moments up to the fourth for the local DOS. How do your results differ from the square two-dimensional lattice?

10.5 Some three-dimensional solids consist of layers stacked one upon another. For such systems, the electronic structure is often dominated by interplanar atomic interactions and the *intra*layer interaction can be ignored. One of the simplest such systems is graphite.

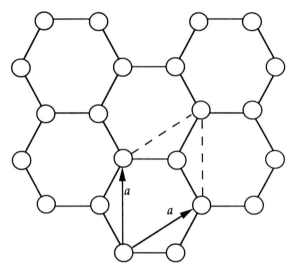

Fig. 10.7. Part of an infinite graphite sheet, showing the unit cell.

Each graphite layer consists of carbon atoms with three neighbors each so as to constitute a hexagonal structure, as shown in Fig. 10.7. There are four electrons in each carbon atom which can be thought of as being distributed equally between the 1s and three 2p orbitals. σ bonds are formed between the s orbital and the p_x and p_y orbitals within the plane. As a result, there is a σ-bonding band and an empty σ-antibonding band.

The fourth valence electron, in the p_z orbital perpendicular to the plane, gives rise to the interesting electronic structure of graphite. This electron participates in π bonding.

The unit cell of a graphite plane contains two atoms, as shown in Fig. 10.7. The two atoms are equivalent by symmetry, but they are not equivalent by the translational operations that make up the lattice.

(a) Write down a general expression for the Bloch wavefunction for the p_z electrons. Note that the wavefunction must have a contribution from each of the two atoms in a unit cell. Draw schematic diagrams for the wavefunctions at three special $\vec{k}$ points: Γ $\vec{k} = (0,0)$; Q $\vec{k} = (2\pi/\sqrt{3}a, 0)$; P $\vec{k} = (2\pi/\sqrt{3}a, 2\pi/3a)$. Note that for each $\vec{k}$ value, there are two possible combinations. For each sketch, decide whether the wavefunction is predominantly bonding, predominantly antibonding, or nonbonding.

(b) Calculate the band structure for the p_z electrons. First find a general expression for $\mathscr{E}(\vec{k})$ and then find explicit values at the Γ, P, and Q points. Plot the band structure around the complete circuit $\Gamma \to Q \to P \to \Gamma$. Be certain to check that your band structure is consistent with your wavefunctions.

10.6 Consider a face-centered-cubic crystal with one atom per lattice site and lattice constant a. The valence state of each atom is an s state. The on-site Hamiltonian matrix elements are α and the hopping integrals between nearest neighbors are β ($\beta < 0$). All other Hamiltonian matrix elements are zero.

(a) Show that the band structure is given by:

$$\mathscr{E}(\vec{k}) = \alpha + 4\beta[\cos(k_x a)\cos(k_y a) + \cos(k_y a)\cos(k_z a) + \cos(k_z a)\cos(k_x a)] \tag{P10.1}$$

The reciprocal space lattice vectors for the fcc lattice are found using Eqs. 8.7.

(b) In the limit of small band fillings show that $\mathscr{E}(\vec{k})$ becomes:

$$\mathscr{E}(\vec{k}) = \alpha + 12\beta - 4\beta a^2 k^2 \tag{P10.2}$$

What is the functional form for the DOS in the limit of small band fillings?

(c) The Brillouin zone is constructed by the Wigner–Seitz method. Using the reciprocal lattice constructed above, choose one lattice point and draw vectors between it and all of its surrounding atoms (not just neighbors). Then divide each of these vectors by a plane perpendicular to the vector and halfway between the atom and those surrounding it. The volume included by all of these planes is the Wigner-Seitz first Brillouin zone. You will find that the Brillouin zone for the fcc lattice is a bcc lattice. As the Brillouin zone is populated with more electrons, what are the last points of the zone to be filled?

(d) What are the maximum and minimum values of $\mathscr{E}(\vec{k})$?

(e) Using the moments theorem, calculate the first three moments of the local DOS and sketch what you think the local DOS should look like throughout the entire band width. Compare with what you have found above.

10.7 Exercise 10.3.

10.8 Exercise 10.4.

10.9 Exercise 10.5.

11

Crystal structures

11.1 Introduction

Few things are more exquisite than the symmetry and perfection manifested by a crystal. It is remarkable to hold a near-perfect cube of pyrite (FeS_2) or a piece of quartz (SiO_2) with its large hexagonal 'fingers' and realize that these crystals consist of a huge number of *atoms*. Some quartz crystals can weigh up to 130 kg and be about a meter in length! Through visits to museums or our everyday experience, most of us have some intuitive idea about what constitutes a crystal structure. But can we proffer a guess as to how many different crystal structures there are?

The answer is, in principle, infinitely many! Of course, in practice only a finite number of crystal structures have been discovered. We will discuss how to describe these crystal structures mathematically, but we will not go into how one might determine them experimentally, as that is the subject of an entire book by itself.

Probably everyone has seen the 'equation' for a crystal structure,

$$\text{crystal structure} = \text{lattice} + \text{basis} \tag{11.1}$$

But what does this really mean? Let us examine each piece of Eq. 11.1 systematically.

A lattice is an infinite array of points in space, in which each point has identical surroundings. A lattice can be generated, or described, by a set of vectors

$$\vec{t}_n = n_1\vec{a} + n_2\vec{b} + n_3\vec{c} \tag{11.2}$$

where n_1, n_2, and n_3 are any integers and $\vec{a}$, $\vec{b}$, and $\vec{c}$ are vectors with $\vec{a}$ and $\vec{b}$ noncollinear and $\vec{c}$ not in the $\vec{a}$–$\vec{b}$ plane. The lattice is the set of points at the end of all of these vectors. This highlights an important point: *a lattice is*

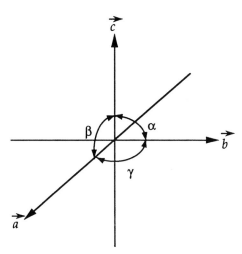

Fig. 11.1. Coordinate system showing the axes $\vec{a}$, $\vec{b}$, and $\vec{c}$ and the interaxial angles α, β, and γ.

just a mathematical construct! Atoms do not necessarily coincide with lattice points and the two concepts cannot be used interchangeably.

The volume in space $\vec{a} \cdot (\vec{b} \times \vec{c})$ will fill all space when translated by $\vec{t}_n$. This volume is called a *unit cell*. If it contains just one lattice point, then it is a *primitive unit cell*. If it contains more than one lattice point, then it is called *nonprimitive* or sometimes multiply primitive. How a primitive unit cell is chosen is not unique, but the volume of each primitive unit cell is the same.

We have not placed any restrictions on the vectors $\vec{a}$, $\vec{b}$, and $\vec{c}$ which define the unit cell. If we use symmetry operations to do so, then we can generate the seven crystal systems. Before examining a specific example, note a convention. Figure 11.1 shows a right-handed set of axes $\{\vec{a}, \vec{b}, \vec{c}\}$ with interaxial angles $\{\alpha, \beta, \gamma\}$. The seven crystal systems are the crudest manner in which we can classify crystal structures. These crystal systems arise from applying proper and improper rotations to the unit-cell axes or translation vectors of the lattice. The rotations impose certain geometrical constraints on the relationships between the unit-cell axes. We have already noted in Ch. 2 that only proper rotations C_n and improper rotations S_n with $n=1, 2, 3, 4, 6$ are allowed. Five-fold rotations cannot fill space properly.

Let us look at the specific example of the triclinic crystal system to see how restrictions are placed on the axes and interaxial angles. The symmetry restriction is that the unit cell contain only the identity operation $\{E\}$

or, alternatively, that it contain only the inversion $\{i\}$. Now apply these operations to an *arbitrary* point in the unit cell $\vec{r} = x\vec{a} + y\vec{b} + z\vec{c}$, where $x, y, z \neq integers$.

$$E\vec{r} = \vec{r'} = \begin{pmatrix} 1 & 0 & 0 \\ 0 & 1 & 0 \\ 0 & 0 & 1 \end{pmatrix} \begin{pmatrix} x \\ y \\ z \end{pmatrix} = x\vec{a} + y\vec{b} + z\vec{c}$$
$$= x'\vec{a} + y'\vec{b} + z'\vec{c}$$

(11.3)

Or, alternatively,

$$i\vec{r} = \vec{r'} = \begin{pmatrix} -1 & 0 & 0 \\ 0 & -1 & 0 \\ 0 & 0 & -1 \end{pmatrix} \begin{pmatrix} x \\ y \\ z \end{pmatrix} = -x\vec{a} - y\vec{b} - z\vec{c}$$
$$= x'\vec{a} + y'\vec{b} + z'\vec{c}$$

(11.4)

For both of these symmetry operations, there are *no* special relations imposed on the axes. Therefore,

$$a \neq b \neq c$$
$$\alpha \neq \beta \neq \gamma$$

and for the triclinic crystal system, the unit cell lengths and interaxial angles can take on any values, independent of one another. Experimentally, it may be found that some quantities are equal. This, however, is a question of precision and not one of theory.

We can continue adding other and more symmetry operations to obtain the restrictions on the other crystal systems. A complete list of the seven crystal systems is given in Table 11.1 and the primitive unit cells are shown in Fig. 11.2. Notice that there are two 'settings' for the monoclinic crystal system: in the 1st setting, the interaxial angles α and β are equal to 90°, whereas in the 2nd setting, the interaxial angles α and γ are equal to 90°. For either setting, the remaining interaxial angle is greater than 90°.

11.2 Bravais lattices

Now that we know how to obtain the seven crystal systems, we might think that, by assigning a primitive lattice to each crystal system, we would end up with 7 space lattices. Moritz Frankenheim was the first person to describe and enumerate the complete set of space lattices systematically. He found

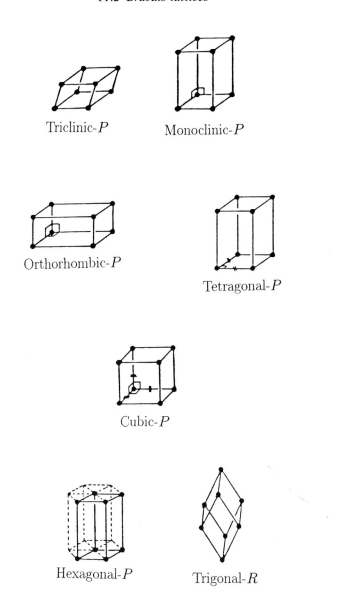

Fig. 11.2. Primitive unit cells of the seven crystal systems.

Table 11.1. *Table of the seven crystal systems showing the symmetry operation(s) used to generate them and the constraints placed on the unit-cell axes and interaxial angles*

Symmetry operation(s)	Crystal system	Constraints
E or i	Triclinic	No conditions
C_2 or σ	Monoclinic	$\alpha = \beta = 90°$
		(1st setting)
		$\alpha = \gamma = 90°$
		(2nd setting)
3 two-fold axes	Orthorhombic	$\alpha = \beta = \gamma = 90°$
$4(C_4)$	Tetragonal	$a = b$
		$\alpha = \beta = \gamma = 90°$
4 three-fold axes	Cubic	$a = b = c$
		$\alpha = \beta = \gamma = 90°$
$6(C_6)$	Hexagonal	$a = b$
		$\alpha = \beta = 90°; \gamma = 120°$
$3(C_3)$	Trigonal	Same as hexagonal
	(Rhombohedral)	$(a = b = c; \alpha = \beta = \gamma)$

15 in all. Unfortunately, the trigonal and hexagonal lattices end up being identical and Frankenheim did not notice this until eight years after Bravais (1848) had shown that the correct number is actually 14.

If we assign a primitive lattice to each crystal system and recognize that two are identical, then we end up with only six. Where do the eight other lattices come from? The answer is that besides the six *primitive* lattices (denoted P lattices), we can generate additional lattices by adding lattice points to judicious places in the primitive lattices. This process is called 'centering'. When centering is carried out on a P lattice, we must ask two questions: (1) Is the new lattice still a lattice? and (2) Is the new lattice new?

Within one crystal system, then, we may have a variety of space lattices, denoted P (primitive), I (body-centered), F (all faces centered), or C (only one face is centered). Generally these unit cells contain more than one lattice point per unit cell, although a unit cell can be found with just one lattice point. Very briefly, the centering process is carried out by asking: Can we add additional lattice points to the primitive lattices such that the constraints imposed by symmetry are still maintained? An example will make this clearer.

Consider body centering in a primitive cubic lattice, in which an additional lattice point is added at the point $(\frac{1}{2}, \frac{1}{2}, \frac{1}{2})a$ in each primitive unit cell. Is this new I lattice (from the German 'Innenzentrierung'), shown in Fig. 11.3,

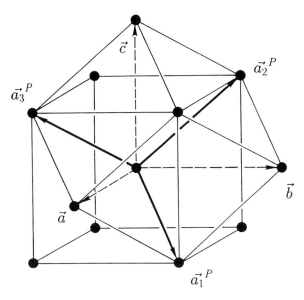

Fig. 11.3. A single unit cell of the centered cubic-*P* lattice (cubic-*I* lattice). $\vec{a}_1^{\,P}, \vec{a}_2^{\,P}$, and $\vec{a}_3^{\,P}$ are the lattice vectors of the cubic-*I* primitive lattice, while $\vec{a}, \vec{b}$, and $\vec{c}$ are the lattice vectors of the multiply primitive cubic-*I* lattice.

still a lattice? Examination of Fig. 11.3 verifies that, indeed, each point in the *I* lattice has identical surroundings. There are now, however, two lattice points per unit cell, one at $(0,0,0)$ and one at $(\frac{1}{2}, \frac{1}{2}, \frac{1}{2})a$. These points possess a different set of symmetry operations from those of lattice points in the primitive cubic lattice, so that the new lattice is indeed new.

Exercise 11.1   Find the symmetry operations associated with a lattice point in the cubic-*P* lattice and in the cubic-*I* lattice.

There are two types of face centering. We may center all faces, in which case the lattice is denoted an *F* lattice, or we may center just one face, in which case the lattice is given the letter corresponding to the centered face. In the cubic crystal system, we can only generate an *F* lattice. However, in the orthorhombic crystal system, we can carry out *F* centering as well as *C* centering, in which only the *C* faces are centered. The conventional unit cells for the 14 Bravais lattices are shown in Fig. 11.4.

Why are we so concerned about whether a unit cell is primitive or nonprimitive? Within the cubic crystal system, the *P* lattice contains just one lattice point per unit cell, the *I* lattice contains two lattice points per unit cell, and the *F* lattice contains four lattice points per unit cell. Thus,

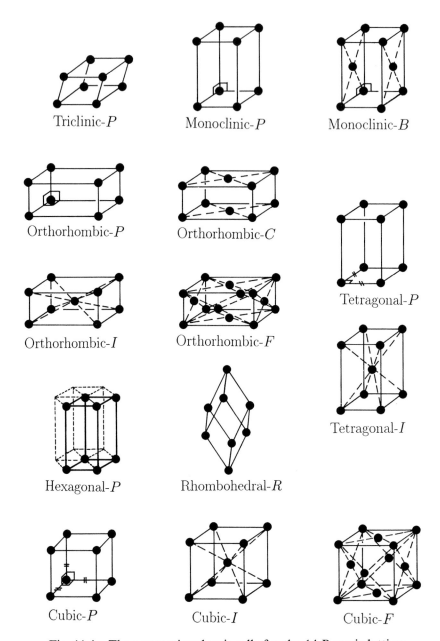

Fig. 11.4. The conventional unit cells for the 14 Bravais lattices.

the *conventional* unit cell for the I lattice has twice the volume of a *primitive* unit cell and the F lattice has four times the volume of a *primitive* unit cell. For any given conventional unit cell, a corresponding primitive unit cell can be constructed. *But the conventional unit cell displays the full rotational symmetry of the crystal system, while the primitive unit cell does not.* At the same time, because the primitive unit cell contains just one lattice point, it is the smallest cell that describes the full translational symmetry of the Hamiltonian. Thus, the conventional unit cell is particularly helpful if we want to visualize symmetry, but the primitive unit cell is essential if we want to calculate the correct number of bands. The number of bands per unit cell is just the number of basis orbitals per atom summed over all atoms in the basis.

11.3 Crystal structure

If crystal systems are the crudest way of classifying crystal structures, then division of each crystal system into Bravais lattices is the first refinement. The second, and most detailed, refinement is to assign each crystal a space group. Recall from Ch. 3 that a space group contains all point and translation operations that carry the crystal into itself. Therefore, given the space group, we automatically know the Bravais lattice and the crystal system. This relationship is shown in Table 11.2.

But we have digressed long enough about mathematical constructs! Let us return to physical objects. A crystal structure may be obtained by starting with a collection of atoms, called the *basis*, and attaching this collection to every lattice point throughout space with precisely the same orientation. Notice that there need be no relation between the point symmetry (or translational symmetry, if the space group contains glides or screws) of a lattice point and the point symmetry of the atoms constituting the basis. Typically, as we pointed out at the beginning of this chapter, the process of generating a crystal structure can be written as

$$\text{crystal structure} = \text{lattice} + \text{basis} \tag{11.5}$$

but such an 'equation' is confusing because it puts lattice and basis on the same footing. In fact, the lattice and the basis actually exist in two different spaces! The lattice is just a mathematical framework or template, but the basis contains the actual atoms that make up our crystal. It is important not to confuse these two concepts.

A more rigorous way to think about a crystal structure is to say that it is the *convolution* of a basis with a lattice. If $B(\vec{r})$ is a function that defines the

Table 11.2. *Relation between the crystal systems, Bravais lattices, and the 73 symmorphic space groups. The superscript on the Schoenflies space group symbol indicates the particular space group. For more information on the exact nature of the group, see the Bibliography. The R setting of the trigonal crystal system is the rhombohedral system.*

Crystal system	Bravais lattice	Space group
Triclinic	P	$C_1, S_2(C_2)$
Monoclinic	P	$C_2^1, C_{1h}^1, C_{2h}^1$
	B or A	$C_2^3, C_{1h}^3, C_{2h}^3$
Orthorhombic	P	$D_2^1, C_{2v}^1, D_{2h}^1$
	$A, B,$ or C	$D_2^6, C_{2v}^{11}, C_{2v}^{14}, D_{2h}^{19}$
	I	$D_2^8, C_{2v}^{20}, D_{2h}^{25}$
	F	$D_2^7, C_{2v}^{18}, D_{2h}^{23}$
Tetragonal	P	$C_4^1, S_4^1, C_{4h}^1, D_4^1, C_{4v}^1, D_{2d}^1, D_{2d}^5, D_{4h}^1$
	I	$C_4^5, S_4^2, C_{4h}^5, D_4^9, C_{4v}^9, D_{2d}^9, D_{2d}^{11}, D_{4h}^{17}$
Cubic	P	$T^1, T_h^1, O^1, T_d^1, O_h^1$
	I	$T^3, T_h^5, O^5, T_d^3, O_h^9$
	F	$T^2, T_h^3, O^3, T_d^2, O_h^5$
Trigonal	P	$C_3^1, C_{3i}^1, D_3^1, D_3^2, C_{3v}^1, C_{3v}^2, D_{3d}^1, D_{3d}^3$
	R	$C_3^4, C_{3i}^2, D_3^7, C_{3v}^5, D_{3d}^5$
Hexagonal	P	$C_6^1, C_{3h}^1, C_{6h}^1, D_6^1, C_{6v}^1, D_{3h}^1, D_{3h}^3, D_{6h}^1$

basis and $L(\vec{r})$ is a function that defines the lattice, then a crystal structure, described by the function $C(\vec{r})$, arises by convoluting the two functions,

$$C(\vec{r}) = B(\vec{r}) * L(\vec{r}) \tag{11.6}$$

The operation of convolution is denoted by *. As a specific, and perhaps whimsical, example, let $B(\vec{r})$ be a function defining a basis that consists of a shark. If $L(\vec{r})$ is a two-dimensional rectangular lattice, then the crystal structure defined by convoluting $B(\vec{r})$ with $L(\vec{r})$ is the school of sharks depicted in Fig. 11.5.

Now we can understand why there might be an infinity of crystal structures. There are only a finite number of Bravais lattices (14), but there are, in principle, an infinity of bases. We show some of the more common crystal structures in Figs. 11.6–11.10.

Problems

11.1   We have discussed three-dimensional point groups earlier in the text and learned that they are 32 in number. (See Ch. 2.) We also

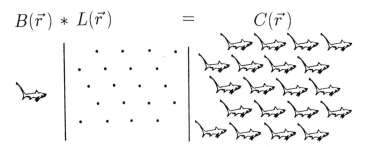

Fig. 11.5. The 'crystal structure' of sharks obtained by convoluting the basis function $B(\vec{r})$ (shark) and the lattice function $L(\vec{r})$ (two-dimensional rectangular lattice). From G. Burns and A.M. Glazer, *Space Groups for Solid State Scientists*, 2nd edition (Academic Press, Boston, 1990), p. 132. Used with permission.

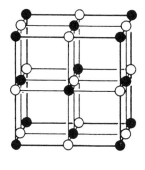

NaCl

Fig. 11.6. The sodium chloride crystal structure. The space lattice is fcc. The basis consists of a Cl^- ion at $(0,0,0)$ and a Na^+ ion at $(\frac{1}{2}, \frac{1}{2}, \frac{1}{2})$. The figure shows one conventional cubic cell.

touched briefly on the 230 space groups – 73 symmorphic (*i.e.*, no glides or screws) and the remainder nonsymmorphic (with glides and/or screws). For studying surfaces and surface-related problems like adsorption, it is important to recognize that there are also two-dimensional point and space groups.

(a) List all the possible types of *plane* symmetry elements (*i.e.*, axes, lines, *etc.*) remembering that you must tile the plane with objects of the designated symmetry. Be specific about the order of the rotation axes.

(b) To generate the point groups in three dimensions, we started with an object that possessed only the identity operation and successively added more symmetry elements. Do the same thing with your symmetry elements in two dimensions to generate the 10 plane point groups.

(c) In three dimensions, we know that there are 14 Bravais lattices. In

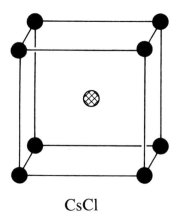

CsCl

Fig. 11.7.   The cesium chloride crystal structure. The space lattice is simple cubic. The basis consists of a Cs^+ ion at $(0,0,0)$ and a Cl^- ion at $(\frac{1}{2},\frac{1}{2},\frac{1}{2})$.

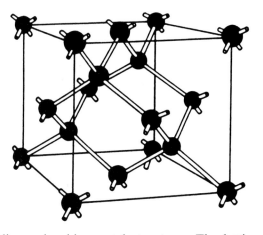

Fig. 11.8.   The diamond cubic crystal structure. The lattice is fcc. The basis consists of two atoms: one at $(0,0,0)$ and another at $(\frac{1}{4},\frac{1}{4},\frac{1}{4})$. The figure shows one conventional cubic unit cell.

two dimensions, there are only 5 plane lattices. Draw a picture of each of these lattices.

(d) By combining the 10 plane point groups with the 5 plane lattices, generate the 13 two-dimensional space groups. It helps immensely to draw pictures.

(e) As with three-dimensional space groups, there exist nonpoint operations in two dimensions. By examining the drawing in Fig. 11.11 by

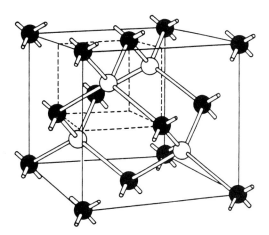

Fig. 11.9. The cubic zinc sulfide crystal structure. The lattice is fcc. The basis consists of an atom A at (0,0,0) and an atom B at $(\frac{1}{4}, \frac{1}{4}, \frac{1}{4})$. The figure shows one conventional cubic unit cell.

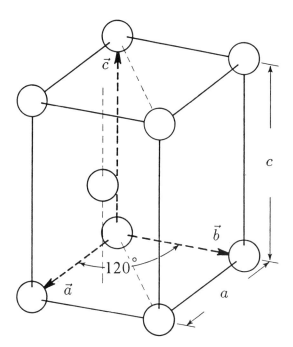

Fig. 11.10. The hexagonal close-packed crystal structure. The space lattice is simple hexagonal. The basis consists of two identical atoms at (0,0,0) and $(\frac{2}{3}, \frac{1}{3}, \frac{1}{2})$.

Fig. 11.11. From C.H. MacGillavry, *Symmetry Aspects of M.C. Escher's Drawings*, 2nd edition (Published for the International Union of Crystallography by A. Oosthoek's Uitgeversmaatschappij NV, Utrecht, 1965), p. 9. Used by permission. © 1994 M.C. Escher/Cordon Art – Baarn – Holland. All rights reserved.

Escher, can you discover what the plane nonpoint operation is? Use a sketch to demonstrate this clearly.

(f) Taking into account the additional nonsymmorphic operation from part (e), an additional 4 plane space groups can be generated, making a total of 17 plane space groups. Find the additional ones.

11.2 The drawings in Fig. 11.12 show patterns of points distributed in orthorhombic unit cells. To which – if any – of the orthorhombic Bravais lattices, P, C, I, or F, does each pattern belong?

11.3 Some common mistakes in crystallography are outlined below. What is wrong with each statement?

(a) Silicon has a diamond lattice.

(b) The CsCl structure is body-centered cubic.

(c) All cubic crystals must contain four-fold rotation axes.

(d) The diamond structure consists of two interpenetrating fcc lattices.

(e) Cu has an fcc lattice with a one-atom basis. Therefore, lattice points and atoms are interchangeable.

11.4 (a) Show that the volume of a unit cell is given by

$$V = abc(1 - \cos^2 \alpha - \cos^2 \beta - \cos^2 \gamma + 2 \cos \alpha \cos \beta \cos \gamma)^{\frac{1}{2}} \quad \text{(P11.1)}$$

(b) Prove that in a given lattice all primitive unit cells are equal in volume, irrespective of the choice of axes.

(a)

(b)

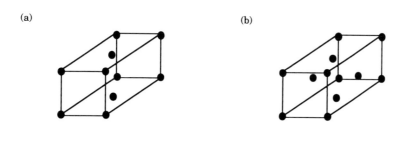

(c)

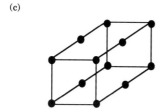

Fig. 11.12.   Patterns of points distributed in orthorhombic unit cells. See Problem 11.2.

       (c) Draw a cubic unit cell with lattice points on all corners. Add a lattice point to the center of each unit-cell face. Show, by shifting the origin of the unit cell, that there are four lattice points in this unit cell.

11.5   (a) Show that every lattice point is at a center of inversion and that midway between the points there is another center of inversion.

      (b) If there is a two-fold rotation axis at every lattice point, show that halfway between neighboring lattice points, another two-fold axis is generated. Do the same for mirror planes.

12

More on bands

12.1 Introduction

We have spent quite a lot of effort trying to solve the Schrödinger equation for simple one-, two-, and three-dimensional systems to generate band structures. One can well appreciate the difficulty of this task for any *real* crystal. To circumvent some of this difficulty, we have introduced concepts like the density of states and the moments theorem. These are powerful concepts not only because they allow us to bypass complicated calculations, but also because they allow us to examine properties of systems for which there is no translational symmetry.

Nonetheless, a great deal can be learned from the band structure of a periodic system. In this chapter, we will learn to navigate our way through a band structure diagram.

12.2 Electrical conductivity

For any system with translational symmetry, we can show (see Problems, Ch. 8) that the group velocity of an electron in a band is

$$\vec{v}_{\mathrm{g}}(\vec{k}) = \frac{1}{\hbar}\nabla_k \mathscr{E}(\vec{k}) \tag{12.1}$$

where the derivative is taken with respect to the wavevector. This equation tells us a number of interesting things without any knowledge of the specific functional form for the band structure. First of all, since the energy is an even function of $\vec{k}$, eigenstates with equal and opposite wavevectors have equal and opposite velocities. Secondly, wherever the slope of the band structure goes to zero, the electron velocity goes to zero for those $\vec{k}$ values. Finally, Eq. 12.1 is independent of time! This means that as an electron travels through a system with a periodic potential, its velocity does not change.

How can it be that, even though the electron interacts with the atoms in the crystal, there is no energy dissipated? What's more disconcerting is if we think of the consequences of a time-independent velocity. Namely, the conductivity of the material is infinite! The answer to this puzzle is really quite simple: electrons are particles, but they also have wave-like behavior. Therefore, an electron can propagate through a perfect crystal without attenuation because the waves can add coherently after scattering off of the atoms.

The earliest classification of the chemical elements was into metals and nonmetals. One of the major distinctions between the two classes, besides appearance, is their electrical conductivity. Metals are characterized by conductivities on the order of $10^5(\Omega\text{-cm})^{-1}$ at room temperature, whereas nonmetals may have conductivities many orders of magnitude smaller. Still, the conductivity of real metals is far from being infinite. This is because real metals have defects, like vacancies or impurities, which do not allow for coherent scattering. In addition, finite temperatures mean that there are lattice vibrations which cause scattering. Scattering by vibrations increases strongly with temperature, but static defect scattering does not change appreciably with temperature.

Let us consider in more detail the conductivity of electrons in our simple one-dimensional H ring. We reproduce the band structure of this system in Fig. 12.1. Fig. 12.1(a) shows a case in which the entire band is full. The Fermi level is at the top of the band and the Fermi wavevector lies at the zone boundaries, $|k_F| = \pm\pi/a$. Now, because there are as many electrons with $0 \leq k \leq \pi/a$ as there are electrons with $0 \geq k \geq -\pi/a$, there are equal numbers of electrons with equal and opposite velocities. Thus, the ground state of this solid has no net motion of electrons and, hence, no electric current flowing. Such a material is considered an insulator. Recall that there are N atoms in our infinite ring and thus a system with a completely filled band must have $2N$ total electrons. Or, each atom contributes two electrons. Although we have examined systems so far with only one atom per unit cell, it can be shown more generally that if a material is an insulator, it has an even number of electrons per unit cell. The converse, however, is not true. An even number of electrons per unit cell does not guarantee that a material is an insulator.

On the other hand, Fig. 12.1(b) shows a partially filled band such that the Fermi wavevector lies somewhere between $k = 0$ and the zone boundary. It is important to recall that because N is large, and the spacing between

(a)

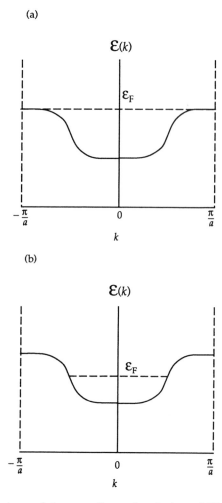

(b)

Fig. 12.1.   Band structure of the one-dimensional ring of H atoms. (a) The band is full, so that the Fermi level is at the top of the band and $k_F = \pm\pi/a$. (b) The band is partially full, so that k_F is somewhere between 0 and the zone boundary.

states in k space is $2\pi/Na$, there are *empty* states very close in energy to the occupied states at k_F.

Exercise 12.1   Using the analytical form for the band energy $\mathscr{E}(k) = \alpha + 2\beta\cos(ka)$, calculate the energy necessary to move an electron at $k_F = \pi/2a$ to the next-highest level. Assume $|\beta| = 1$ eV and $a = 1$ Å. Take N to be any sufficiently large integer or try several different values.

For a small energy input to the crystal, electrons can move to empty bands and, hence, the material is a conductor.

But where does this energy come from? One instance occurs when we heat the crystal. In that case, the electron occupations follow the Fermi distribution function, to which we will return in Ch. 14. Another way is to apply an external electric field, $\vec{E}$. The work $d\mathscr{E}$ done by a *classical* particle traveling a distance $v_g dt$ under the action of a force $e|\vec{E}|$ is

$$d\mathscr{E} = e|\vec{E}|v_g dt \tag{12.2}$$

which becomes

$$d\mathscr{E} = e|\vec{E}|\frac{1}{\hbar}\frac{\partial \mathscr{E}}{\partial k}dt \tag{12.3}$$

after substituting Eq. 12.1 for a one-dimensional system. As always, the results can be readily generalized to two and three dimensions. Consider, then, the acceleration derived from Eq. 12.1,

$$\begin{aligned}
\frac{dv_g}{dt} &= \frac{1}{\hbar}\frac{d}{dt}\frac{\partial \mathscr{E}}{\partial k} \\
&= \frac{1}{\hbar}\frac{\partial^2 \mathscr{E}}{\partial k^2}\frac{dk}{dt}
\end{aligned} \tag{12.4}$$

From Eq. 12.3, we find that $dk/dt = e|\vec{E}|/\hbar$, so that

$$\frac{dv_g}{dt} = \frac{e|\vec{E}|}{\hbar^2}\frac{\partial^2 \mathscr{E}}{\partial k^2} \tag{12.5}$$

Equation 12.5 is the acceleration of just one electron. If we multiply the individual electron acceleration by the electric charge e and the electron density n, and sum over all electrons, we obtain the time derivative of the electric current density,

$$\begin{aligned}
\frac{dJ}{dt} &= \sum_{\text{all electrons}} \frac{d}{dt}(nev_g) \\
&= \frac{e^2}{\hbar^2}|\vec{E}| \sum_{\text{all electrons}} \frac{\partial^2 \mathscr{E}}{\partial k^2}
\end{aligned} \tag{12.6}$$

For a large number of electrons in the system, the sum can be replaced by an integral, so that

$$\frac{dJ}{dt} = \frac{e^2}{\hbar^2}|\vec{E}|\frac{1}{\pi}\int\frac{d^2\mathscr{E}}{dk^2}dk \tag{12.7}$$

Now, imagine that our system consists of a number density of 'effective' electrons, N_{eff}, which are *noninteracting* and *free*. By free we mean not in the constraints of a periodic potential. For a system of noninteracting and free electrons of density N, we have

$$\frac{dJ}{dt} = \frac{e^2 |\vec{E}|}{m} N \tag{12.8}$$

Comparing Eq. 12.8 with Eq. 12.7, we find

$$N_{\text{eff}} = \frac{1}{\pi} \frac{m}{\hbar^2} \int \frac{d^2 \mathscr{E}}{dk^2} \, dk \tag{12.9}$$

Equation 12.9 is only strictly valid at $T = 0$ K, because we have implicitly assumed that the probability of occupation is a step function. Returning to Fig. 12.1(b), we can see that for a Fermi energy somewhere in the middle of the band, the effective number of conduction electrons is

$$
\begin{aligned}
N_{\text{eff}} &= \frac{1}{\pi} \frac{m}{\hbar^2} \left[\left(\frac{d\mathscr{E}}{dk} \right)_{k=k_{\text{F}}} - \left(\frac{d\mathscr{E}}{dk} \right)_{k=-k_{\text{F}}} \right] \\
&= \frac{2}{\pi} \frac{m}{\hbar^2} \left(\frac{d\mathscr{E}}{dk} \right)_{k=k_{\text{F}}} \tag{12.10}
\end{aligned}
$$

Therefore, *the conductivity is proportional to the slope of the band structure at the Fermi energy.* Everywhere in the middle of the band this is nonzero, so that we have a finite conductivity only for partially filled bands. As an aside, notice also that, for a completely filled band, $k_{\text{F}} = \pm \pi/a$ and the slope vanishes for our simple H ring, a result that we anticipated already. It should be stressed that this model is semi-classical in its treatment of the conductivity, because Eq. 12.2 is strictly only valid for classical particles.

Closer examination of Eq. 12.10 reveals a few problems as well. Specifically, for the H ring model, $\mathscr{E}(k) = \alpha + 2\beta \cos(ka)$, so that $d\mathscr{E}/dk = -2\beta a \sin(ka) = 2|\beta| \sin(ka)$. But then the conductivity is negative for some values of k! We turn to this problem next.

12.3 The effective mass

The mass of a free, nonrelativistic electron is a well-defined quantity. But what happens when electrons are subjected to the periodic potential of a crystal? Comparing Eq. 12.5 with what we would expect for a free particle,

$$m \frac{dv}{dt} = e|\vec{E}| \tag{12.11}$$

we can define the *effective mass* of an electron as

$$m^* \equiv \hbar^2 \left(\frac{\partial^2 \mathscr{E}}{\partial k^2} \right)^{-1} \tag{12.12}$$

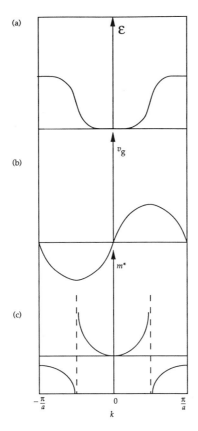

Fig. 12.2. A composite plot showing (a) the band structure, (b) the group velocity, and (c) the effective mass of the H ring.

Plots of the band structure, the group velocity, and the effective mass for the H ring are shown in Fig. 12.2.

Even for this very simple model, there are some very strange effects. From Fig. 12.2(c), in particular, we see that the effective mass is positive for k in the vicinity of zero. As k increases toward $\pm\pi/2a$, m^* diverges to infinity. Furthermore, for $|k| > \pi/2a$, m^* is actually less than zero! As we have an unshakeable belief that all physical systems ought to follow rational rules, what does it mean to have a negative effective mass?

Consider Eq. 12.6 for the time derivative of the current density,

$$\frac{dJ}{dt} = e^2|\vec{E}| \sum_i \frac{1}{m_i^*} \tag{12.13}$$

where we have substituted Eq. 12.12 and the index i runs over all electrons.

For a full band,

$$\frac{dJ}{dt} = e^2 |\vec{E}| \sum_i \frac{1}{m_i^*} = 0 \tag{12.14}$$

Assume now that we can reach into the band and selectively remove one electron. In this case,

$$\frac{dJ_h}{dt} = e^2 E \sum_{i, i \neq j} \frac{1}{m_i^*} \tag{12.15}$$

where j is the label for the missing electron and we have added the subscript 'h' to the current density. We will explain this latter subscript below. For the missing electron,

$$\frac{dJ}{dt} = \frac{e^2 E}{m_j^*} \tag{12.16}$$

Adding Eqs. 12.15 and 12.16, and taking Eq. 12.14 into account,

$$e^2 |\vec{E}| \left(\frac{1}{m_j^*} + \sum_{i, i \neq j} \frac{1}{m_i^*} \right) = 0 \tag{12.17}$$

Therefore,

$$\frac{dJ_h}{dt} = -e^2 E \frac{1}{m_j^*} \tag{12.18}$$

As we saw in the previous section, though, the electron effective mass in the upper part of the band is negative. Thus,

$$\frac{dJ_h}{dt} = \frac{e^2 E}{|m_j^*|} \tag{12.19}$$

so that *an electron missing from the top of the band is exactly identical with one electron present at the bottom of an otherwise empty band.*

Think carefully now. A missing electron in a full band is analogous to a vacancy in a crystal. We can focus on the motion of the atoms surrounding the vacant site, or we can focus on the motion of the vacancy. Here, too, we can focus on the dynamics of the missing electron. In that case, this *hole* moves under the influence of an external field as though it had a positive charge. The conductivity does not become negative – it is simply reinterpreted as being carried by positively charged holes that move in the direction opposite to the motion of electrons. Hence, the subscript h in Eq. 12.15. In all respects, the fictitious holes can be regarded as electrons with the opposite sign.

12.4 Symmetry constraints on bands

In discussing conductivity, we have made use of the observation that the energy eigenvalues of our one-dimensional H ring are symmetric about $k = 0$. What happens, however, if we choose a reciprocal lattice vector that is outside of the first Brillouin zone, *i.e.*, $|k| > \pi/a$? Consider the one-dimensional statement of Bloch's theorem: For a potential of the form $V(x) = V(x + a)$, the wavefunction can always be written as the product $\psi_k(x) = e^{ikx}u(x)$, where $u(x) = u(x + a)$. If we choose a wavevector outside of the first Brillouin zone, k', then a reciprocal lattice vector, G, can always be found to shift k' back into the first zone,

$$
\begin{aligned}
\psi_{k'}(x) &= e^{ik'x}u_{k'}(x) \\
&= e^{i(k-G)x}u_{k'}(x) \\
&= e^{ikx}\left[e^{-iGx}u_{k'}(x)\right] \\
&= e^{ikx}u_k(x) \qquad (12.20)
\end{aligned}
$$

where $u_k(x) \equiv e^{-iGx}u_{k'}(x)$. Therefore, $\psi_{k'}(x)$ and $\psi_k(x)$ have the *same* eigenvalue, and we need only solve the Schrödinger equation in the first Brillouin zone. Restricting our choice of k to the first Brillouin zone is called working in the *reduced zone* scheme.

Translational symmetry severely reduces the scope of the unique energy eigenvalues. What are the consequences of point symmetry operations? Consider the simple H ring once again. Choose two points in the Brillouin zone k_1 and k_2 such that $k_2 = k_1 - 2\pi/a$, as shown in Fig. 12.3. The two boundary points are related by a mirror plane perpendicular to the k axis at $k = 0$. Now,

$$
\mathcal{E}(k_1) = \mathcal{E}(-k_1) = \mathcal{E}(k_2) \qquad (12.21)
$$

Let k_1 approach π/a so that at the same time k_2 approaches $-\pi/a$,

$$
\lim_{h\to+0} \mathcal{E}\left(\frac{\pi}{a} - h\right) = \lim_{h\to+0} \mathcal{E}\left(-\frac{\pi}{a} + h\right) = \lim_{h\to+0} \mathcal{E}\left(-\frac{\pi}{a} - h\right) \qquad (12.22)
$$

where the limit is taken so that h approaches zero from the positive side. Equation 12.22 implies that the energy is an even function at the points $\pm\pi/a$. Now, the derivative of $\mathcal{E}$ with respect to k is defined as

$$
\frac{\partial \mathcal{E}(k)}{\partial k} \equiv \lim_{\delta\to 0} \frac{\mathcal{E}(k+\delta) - \mathcal{E}(k-\delta)}{2\delta} \qquad (12.23)
$$

where δ is infinitesimally small. The fact that the energy is an even function at the zone boundary means that the right-hand side of Eq. 12.23 vanishes.

Fig. 12.3.   Schematic diagram of one-dimensional reciprocal space. The zone boundaries are indicated at $\pm\pi/a$. Two wavevectors, k_1 and k_2, are related by a reciprocal lattice vector $2\pi/a$.

If the derivative exists, then it, too, vanishes:

$$\left.\frac{\partial\mathscr{E}(k)}{\partial k}\right|_{\text{zone boundary}} = 0 \qquad (12.24)$$

If the derivative does not exist, then the energy must be cusped at the zone boundary. It can be shown, however, that such discontinuities will not exist. Why?

Before answering this question, let us examine Eq. 12.24 in a more general case. If we move from one to two or three dimensions, then the derivative in Eq. 12.24 becomes the *normal derivative* at a zone boundary that has a reflection operation parallel to it. We write the derivative as the gradient and the normal derivative is the gradient dotted with a unit normal vector, $\hat{n}$,

$$\frac{\partial}{\partial k} \to \hat{n}\cdot\nabla_{\vec{k}} \qquad (12.25)$$

Equation 12.24 can be rewritten as

$$\left.\left(\hat{n}\cdot\nabla_{\vec{k}}\mathscr{E}(\vec{k})\right)\right|_{\text{zone boundary}} = 0 \qquad (12.26)$$

The slope of the band structure vanishes at the zone boundary purely as a consequence of the mirror symmetry about the origin. Cusps cannot exist in the band structure because, from Eq. 12.1, that implies the unphysical situation that the group velocity has a discontinuity.

Exercise 12.2   For the two-dimensional square lattice show that $\hat{n}\cdot\nabla_k\mathscr{E}(\vec{k}) = 0$ for every point along the zone boundary. Note that there are two mirror planes in the square lattice that coincide with the reciprocal space axes. Therefore, we can apply the same argument as above to every point along the zone faces.

12.5 Two-band model

Scattering by defects and lattice vibrations can account for the fact that metals do not have infinite conductivities. Still, we cannot distinguish between metallic and nonmetallic materials within the confines of the simple tight-binding one-band model that we have been discussing so far. We must consider more than one orbital per unit cell, which we will turn to now.

To simplify a complex problem, consider our familiar ring of N hydrogen atoms. Now, however, let us expand the basis set of atomic orbitals on each atom. In Ch. 8, we included only the ground state occupied orbitals; namely, the $1s$ orbitals. This is also known as a *minimal* basis set. An *extended basis set* includes atomic orbitals that are normally unoccupied in the ground state. We will explicitly consider a basis set that consists of the $2s$ and $2p$ orbitals. We assume that the $1s$ orbitals are core orbitals and, hence, do not interact. One unit cell still contains just one atom, so that our crystal consists of N unit cells, but there are now a total of *2N states* in our system. The system can accommodate a total of $4N$ electrons.

The atomic orbitals on any atom j are labeled $|\phi_s(j)\rangle$ and $|\phi_p(j)\rangle$. The on-site Hamiltonian matrix elements for the two basis states are different for symmetry reasons. We label them $\langle\phi_s(j)|\mathcal{H}|\phi_s(j)\rangle \equiv \alpha_s$ and $\langle\phi_p(j)|\mathcal{H}|\phi_p(j)\rangle \equiv \alpha_p$, with $\alpha_p > \alpha_s$. There are three nonzero hopping integrals,

$$\begin{aligned}
\beta_s &\equiv \langle\phi_s(j)|\mathcal{H}|\phi_s(j\pm 1)\rangle \\
\beta_p &\equiv \langle\phi_p(j)|\mathcal{H}|\phi_p(j\pm 1)\rangle \\
\beta_{sp} &\equiv \langle\phi_s(j)|\mathcal{H}|\phi_p(j\pm 1)\rangle
\end{aligned} \tag{12.27}$$

The normalized Bloch wavefunction is a linear combination of the *two* atomic basis orbitals on *each* atom,

$$|\psi_k\rangle = \frac{1}{\sqrt{N}}\sum_j e^{ijka}(c_s(k)|\phi_s(j)\rangle + c_p(k)|\phi_p(j)\rangle) \tag{12.28}$$

where the spacing between adjacent atoms is a. Inserting the wavefunction into the Schrödinger equation gives,

$$\begin{aligned}
&\frac{1}{\sqrt{N}}\sum_j e^{ijka}(c_s(k)\mathcal{H}|\phi_s(j)\rangle + c_p(k)\mathcal{H}|\phi_p(j)\rangle) = \\
&\frac{\mathscr{E}_k}{\sqrt{N}}\sum_j e^{ijka}(c_s(k)|\phi_s(j)\rangle + c_p(k)|\phi_p(j)\rangle)
\end{aligned} \tag{12.29}$$

Taking the overlap of Eq. 12.29 successively with $\langle\phi_s(m)|$ and $\langle\phi_p(m)|$, we obtain two equations,

$$\begin{aligned}
(\alpha_s + 2\beta_s\cos(ka) - \mathscr{E}_k)c_s(k) + 2\beta_{sp}\cos(ka)c_p(k) &= 0 \\
2\beta_{sp}\cos(ka)c_s(k) + (\alpha_p + 2\beta_p\cos(ka) - \mathscr{E}_k)c_p(k) &= 0
\end{aligned} \tag{12.30}$$

Exercise 12.3   Show the last step.

We have used the definitions in Eq. 12.27 above and the orthonormality of the basis functions to obtain Eqs. 12.30. For Eqs. 12.30 to have a nontrivial solution, the determinant of the coefficients must vanish,

$$
\begin{vmatrix}
\alpha_s + 2\beta_s \cos(ka) - \mathscr{E}_k & 2\beta_{sp} \cos(ka) \\
2\beta_{sp} \cos(ka) & \alpha_p + 2\beta_p \cos(ka) - \mathscr{E}_k
\end{vmatrix} = 0
$$

Multiplying out the determinant and solving the resulting quadratic equation gives *two* roots for $\mathscr{E}_k$,

$$
\mathscr{E}_k = \frac{\epsilon_s + \epsilon_p}{2} \pm \left(\frac{(\epsilon_s - \epsilon_p)^2}{4} + 4\beta_{sp}^2 \cos^2(ka) \right)^{\frac{1}{2}} \tag{12.31}
$$

where we have defined

$$
\epsilon_s \equiv \alpha_s + 2\beta_s \cos(ka)
$$
$$
\epsilon_p \equiv \alpha_p + 2\beta_p \cos(ka) \tag{12.32}
$$

Exercise 12.4   Derive Eqs. 12.31.

Note that Eqs. 12.32 are identical with the band structure for the single atomic orbital basis. From Eq. 12.31, then, it is clear that we now have *two* bands, which we will relabel as $\mathscr{E}_k^{(s)}$ and $\mathscr{E}_k^{(p)}$. The sign of the hopping matrix element is different in each case: $\beta_s < 0$, while $\beta_p > 0$. The magnitude is different for each as well. Plots of these functions are shown in Fig. 12.4. The lower-energy band is associated with the minus sign between the two terms, while the higher-energy band is associated with the plus sign between the two terms. The new superscript is called the *band index* and it counts the number of bands per unit cell. As a check, the number of bands per unit cell should be equal to the number of orbitals in the basis set.

If the interaction hopping matrix element, β_{sp}, is set to zero, the two bands decouple and we recover $\mathscr{E}_k^{(s)} = \mathscr{E}_s$ and $\mathscr{E}_k^{(p)} = \mathscr{E}_p$. This consequence shows us that for an extended basis set, the bands are mixtures of all the atomic orbitals.

The total DOS for this two-band system is plotted in Fig. 12.5. The bands are not necessarily of the same width, since the width depends on β_s and β_p, which can differ. Furthermore, there may be a range of energies for which there are no states between the lower band and the upper band. This is marked in Fig. 12.5 as $\mathscr{E}_g$ and is identified as the *band gap* of this material.

We have discussed earlier in this chapter the conductivity of electrons in bands. That discussion can be extended here. If the lower band is only partially occupied, then the material is a metal. If, however, the lower band is filled, then there is a region of zero DOS of magnitude $\mathscr{E}_g$ before there

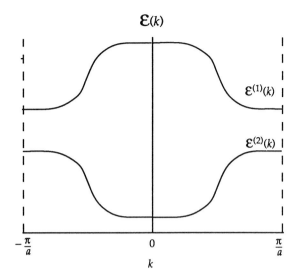

Fig. 12.4. Schematic band structure for the two-band model with a basis of one s and one p orbital.

are more empty states. If $\mathscr{E}_g$ is greater than approximately 2 eV, the material is considered an insulator. A typical example is diamond, for which $\mathscr{E}_g \sim 5$ eV. If $\mathscr{E}_g$ is less than 2 eV, then carriers can be thermally excited across the band gap and the material is a semiconductor. Typical examples include Si, with $\mathscr{E}_g = 1.2$ eV, and Ge, with $\mathscr{E}_g = 0.7$ eV. We will discuss bonding in Si more extensively in Ch. 13.

What determines the magnitude of the band gap? The band gap is defined as the energy difference between the top of the lower band and the bottom of the upper band. This difference depends on the difference between the on-site matrix elements in conjunction with the individual band widths and also on the inter-band hopping matrix element. Note, however, that the existence of a gap *does not* depend on translational symmetry. Rather, it originates from the local chemistry of the constituent atoms.

12.6 The Peierls distortion

Up to this point, we have implicitly or explicitly assumed that $T = 0$ K and so our atoms are truly static. In reality, of course, we cannot neglect the vibrations introduced at finite temperatures. In 1937, Jahn and Teller proved the following theorem:

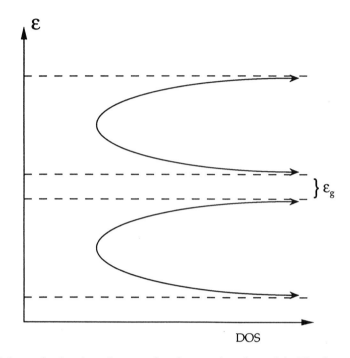

Fig. 12.5. Schematic density of states for the two-band model. The lower contribution is from the *s* basis orbitals and the upper contribution is from the *p* basis orbitals.

Theorem 12.1 *Any nonlinear molecular system in a degenerate electronic state will be unstable and will undergo some kind of distortion that will lower its symmetry and split the degeneracy.*

There are two asides to this theorem. Firstly, the theorem predicts that distortions occur only for degenerate states. Secondly, there is no indication of the geometrical nature of the distortion nor how great it will be.

The Jahn–Teller distortion occurs as a result of coupling between vibrational motion and electronic motion. That is, for a specific normal mode of the system, the molecule is distorted, and the electrons are able to reorganize themselves to lower the system's total energy. As a consequence, electrons become localized.

In the 1950s, Peierls was the first to apply these ideas to solids. Let us return one more time to the H ring. Recall that there is a degeneracy in the band structure of this system of the type $\mathscr{E}(k) = \mathscr{E}(-k)$ for all k throughout the Brillouin zone. Schematic diagrams for three wavefunctions at $k = 0$, $k = \pm\pi/2a$, and $k = \pm\pi/a$ are reproduced in Fig. 12.6. The lowest-energy

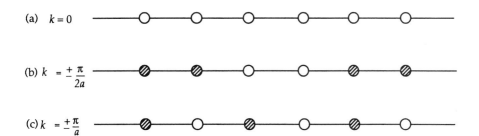

Fig. 12.6. Diagrams for wavefunctions of the H atom ring at (a) $k = 0$; (b) $k = \pm\pi/2a$; and (c) $k = \pm\pi/a$.

Fig. 12.7. The lowest-energy vibrational normal mode for the H atom ring. Atoms execute motion in the direction indicated by the arrows as well as opposite to it.

normal mode, besides a trivial translation, for the chain system is an alternate compression and extension of bonds, as shown in Fig. 12.7. The arrows in this figure indicate only one-half of the atomic motion; of course, all atoms execute an identical distortion in the opposite direction.

What happens to the various bands as the chain undergoes such vibrations? The wavefunction for $k=0$ does not change; since all interactions are bonding, the increase in the system's energy for compressed bonds is exactly cancelled by the decrease for extended bonds. A similar argument can be applied to the $k = \pi/a$ wavefunction. However, the wavefunction at the zone midpoint is different. Coupling the normal mode of Fig. 12.7 with the wavefunction from Fig. 12.6(b) actually moves pairs of bonding atoms closer together while at the same time moving pairs of antibonding atoms farther apart. The degeneracy of the system at $k = \pm\pi/2a$ is broken (as it is, indeed, for other values of k in the vicinity) and, overall, the energy of the system is lowered.

The net impact on the band structure is that a band gap is opened up at $k = \pm\pi/2a$. This is sketched in Fig. 12.8. The magnitude of the gap depends on many things, such as the strength of coupling between atomic and electronic motion.

The Peierls instability is important for determining the structure of many quasi-one-dimensional solids. Polyacetylene is probably one of the most extensively studied solids that undergoes a Peierls distortion. The structure

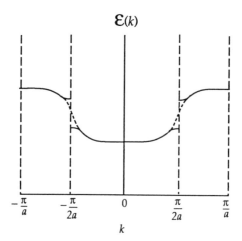

Fig. 12.8. Band structure of the H ring in the first Brillouin zone. Gaps opened up by the Peierls instability are shown for k in the vicinity $\pm\pi/2a$.

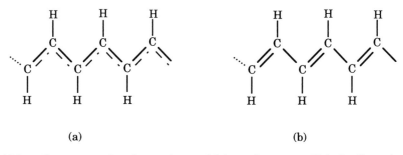

    (a)   (b)

Fig. 12.9.   Structure of polyacetylene which undergoes a Peierls distortion. (a) Structure before distortion showing all π electrons delocalized along the carbon backbone (the dashed lines). (b) Structure after distortion; the π electrons are now delocalized into double bonds which have a shorter bond length.

of polyacetylene before symmetry-lowering is shown in Fig. 12.9(a), while the distorted form is shown in Fig. 12.9(b). One major result of the Peierls distortion is that it localizes the electrons between atoms (Fig. 12.9b) instead of allowing them complete delocalization along the entire carbon chain (Fig. 12.9a).

12.7 Measuring band structures experimentally

We conclude this chapter on bands by considering two questions: Are the band structures that we have calculated correct? How do we know? The

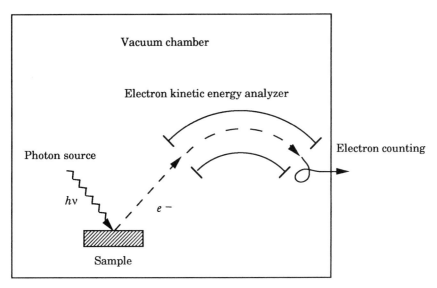

Fig. 12.10. Schematic arrangement of the essential elements of a photoelectron spectrometer. After P.A. Cox, *The Electronic Structure and Chemistry of Solids* (Oxford University Press, Oxford, 1987), p. 26.

most straightforward way of answering these questions is to probe the system experimentally. Spectroscopic methods provide direct information about the electronic structure of solids. We thus turn to a discussion of photoelectron spectroscopy (PES), which is a commonly used technique of investigating the filled levels in solids.

12.7.1 General principles

In PES, a beam of monoenergetic photons of energy $h\nu$ illuminates the sample. Electrons absorb energy and, if they have sufficient energy, are ejected from the sample. A detector measures the kinetic energy ($\mathscr{E}_{KE}$) of these ionized electrons, which is related to their binding energy ($\mathscr{E}_{BE}$) in the solid by

$$\mathscr{E}_{KE} = h\nu - \mathscr{E}_{BE} \tag{12.33}$$

The binding energy reflects directly the band structure. A schematic arrangement of a photoelectron spectrometer is shown in Fig. 12.10.

Depending on the photon energy, the kinetic energy of the emitted electrons can vary from 10 to 1000 eV. Because the electrons scatter strongly inside the solid, there is a background superimposed on the binding energy distribution. Those electrons that scatter less, and, hence, contribute most to the PE spectrum, come from the near-surface region. PES thus probes the electronic structure of the solid within about 10–20 Å of the surface! If there are impurities adsorbed on the surface, they contribute proportionately more than the bulk of the solid. To remedy this problem, PES is performed under ultra-high-vacuum conditions after the sample has been properly cleaned *in situ*.

12.7.2 *Selection rules*

Let us turn now to consider the quantum mechanics of electronic transitions. There are a number of different kinds of transitions which correspond to changes in electric or magnetic dipoles (or higher multipoles) or polarizability tensors. Electric dipole transitions derive their name from the charge distribution between the initial and final states, which is dipolar. An electromagnetic field can couple efficiently to electric dipole transitions such that these transitions are by far the most intense. If one speaks of 'allowed transitions', it is likely that one means 'electric dipole allowed'.

Now, recall from our discussion of electrical conductivity earlier in this chapter that in a filled band in the absence of an electric field, there are as many electrons with $0 < k \leq k_F$ as there are electrons with $-k_F \leq k < 0$. The *total* wavevector of all electrons is zero. Imagine that we excite one electron from, for example, the *s* band of Fig. 12.4 to the *p* band. The wavevector of the excited electron and the hole left behind must be equal and opposite to ensure wavevector (momentum) conservation,

$$\vec{k}_e + \vec{k}_h = 0 \tag{12.34}$$

Equation 12.34 is the selection rule for electric dipole-allowed transitions. Because there is no change in wavevector, such transitions are called *vertical* or *direct* transitions.

For a band structure of the type in Fig. 12.4, the lowest-energy transition is between the top of the *s* band and the bottom of the *p* band. Because these two energies occur at the same *k* point ($k = \pm \pi/a$ in this case), this material is said to have a *direct* band gap. The lowest-energy electronic transition in this system is shown schematically in Fig. 12.11(a). We could imagine a slightly more complex band structure like that of Fig. 12.11(b). Now, the top of the filled band does not lie at the same *k* point as the bottom of the empty band. The lowest-energy transition in this material is an *indirect* transition and, hence, the material has an *indirect band gap*. Momentum must still be conserved so that phonon absorption or emission is required.

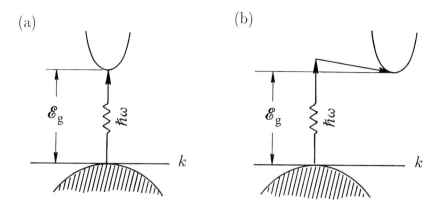

Fig. 12.11. (a) Schematic diagram for the lowest-energy electronic transition for a direct band gap material. (b) Schematic diagram for one of the lowest-energy electronic transitions for an indirect band gap material. The lowest-energy transition from the top of the filled band to the bottom of the empty band does not occur for the same wavevector.

12.7.3 Photoelectron spectroscopy

In a real photoelectron spectroscopy experiment, we have already noted that electrons are ejected from the sample by absorbing energy from incident photons. The final state in the selection rule, Eq. 12.34, is thus a free electron with a plane wave wavefunction. The momentum of a free particle is related to its kinetic energy by

$$p = (2m\mathcal{E}_{KE})^{\frac{1}{2}} \tag{12.35}$$

But electrons will be ejected to all different directions relative to the surface normal. Angular-resolved photoelectron spectroscopy is used to determine the dependence of the spectrum on angle to the surface normal. A schematic diagram is shown in Fig. 12.12.

The detector is placed at some position (θ, ϕ) and the kinetic energy of exiting electrons is determined. From Eq. 12.35 and Fig. 12.12, the wavevector parallel to the surface is given by

$$k_{\parallel} = \frac{1}{\hbar} p \sin \theta$$

$$= \left(\frac{2m\mathcal{E}_{KE}}{\hbar^2} \right)^{\frac{1}{2}} \sin \theta \tag{12.36}$$

If an electron emerges from the sample without scattering, then its wavevector is the same as that of the hole left behind, from the selection rule, Eq. 12.34. As the detector is systematically moved through all angles θ, a spectrum is

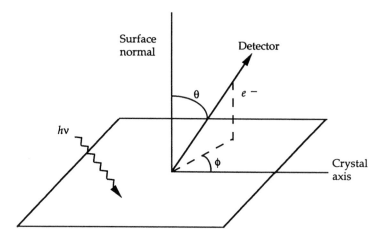

Fig. 12.12.   Geometrical arrangement of an angular-resolved photoelectron spectroscopy experiment. The angle θ is the polar angle from the surface normal and the angle ϕ is the azimuthal angle measured between a crystallographic axis and the projection of the outgoing electron's trajectory. After P.A. Cox, *The Electronic Structure and Chemistry of Solids* (Oxford University Press, Oxford, 1987), p. 118.

gathered which directly reflects the band structure of the solid. By changing the angle ϕ, different directions in k space can be explored.

In the exercises at the end of Ch. 10 we have studied the band structure of a graphite sheet. Let us examine a PES spectrum for graphite to gain a better understanding of the technique. A series of spectra are shown in Fig. 12.13(a) for both normal and off-normal incidence as well as along directions in the plane of the hexagonal Brillouin zone (see inset). These same data, interpreted using Eq. 12.36, are plotted as a band structure in Fig. 12.13(b). The dashed lines in this figure are from theoretical calculations. The σ- and π-derived bands are labeled. Note that the splitting in the π band arises from interplanar interactions, which we did not consider in our simple calculation. Agreement between theory and experiment is quite good, but earlier cautions about cleanliness and surface sensitivity should be kept in mind.

Problems

12.1   Exercise 12.2.

12.2   Consider a two-dimensional square lattice with a one-atom basis. Focus only on in-plane bonding and ignore core s orbitals. The orbitals involved in bonding are therefore p_x and p_y.

(a)

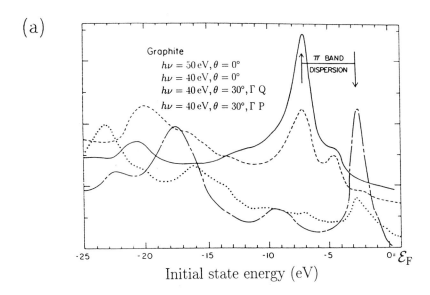

(b)

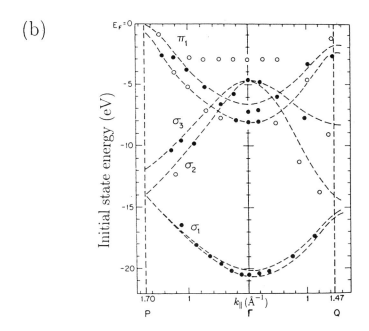

Fig. 12.13. Photoelectron spectra measured at different polar angles and in different k directions are shown in the top figure. The band structure that results from these data is shown as a series of points in the bottom figure. The dashed lines in the bottom figure are from theoretical calculations. From I.T. McGovern *et al.*, *Physica B* **99**, 415 (1980). Used with permission.

(a) Set up tight-binding wavefunctions for this system. Find the secular equations and solve for the eigenvalues at Γ, X, and M.

(b) Plot the band structure around the circuit $\Gamma \to X \to M \to \Gamma$. Be certain to consider the effects of any symmetry elements.

(c) For Γ, X, and M, draw schematic figures for the wavefunctions. Are your figures consistent with the band structure?

(d) Draw a schematic figure for the density of states.

12.3 For the two-band model discussed in the text, find an expression for the effective mass. Plot m^* as a function of k. What are the important variables that determine m^*? How might tiny effective masses arise? Can you suggest ways of engineering materials with tiny m^*?

13

Some case studies in two and three dimensions

13.1 Introduction

In 1954, Slater and Koster† were the first to use the tight-binding method systematically to calculate the band structure for crystals with the simple cubic, face-centered cubic, body-centered cubic, and diamond structures. One of the worrisome aspects to the tight-binding method is how to calculate the Hamiltonian matrix elements. We have seen that the various atomic wavefunctions can be complicated functions and finding explicit expressions for all of the on-site and nearest-neighbor interactions is tedious at best. What's more, how do we know that we have calculated these integrals correctly? What Slater and Koster did was to adjust the Hamiltonian matrix elements to energies obtained from more accurate calculations at special points in the Brillouin zone. Then, it was assumed that the tight-binding parameters so found could be used to examine the band structure throughout all parts of the Brillouin zone. The tight-binding method was thus offered as an interpolative scheme between special k points.

Tight-binding calculations performed later in the 1970s also relied on the idea that the matrix elements were fitted parameters. But the values were taken from spectroscopy studies done on various materials. These semi-empirical calculations offered a great deal of intuitive insight into bonding in various crystals. We will examine several case studies in this chapter that make use of the tight-binding methodology. In addition, we will apply some of our other tools to examine the density of states and the conductivity of these real systems.

† J.C. Slater and G.F. Koster, *Phys. Rev.* **94**, 1498 (1954).

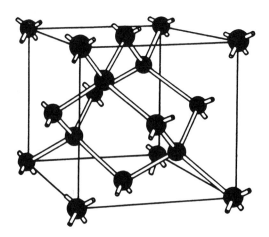

Fig. 13.1. Conventional unit cell of silicon in the diamond structure.

13.2 Silicon in the diamond structure

In calculating the band structure of Si, we combine both of the complications that arose in studying the two-band model (Ch. 12) and the electronic structure of a graphite sheet (Problem 10.5). Namely, we have more than one atomic orbital per atom and we have more than one atom per unit cell. Still, these features only add slight complications, which we are able to deal with, given what we already know.

The electronic structure of a Si atom is $[Ne]3s^2 3p^2$, where the brackets around Ne indicate that the core electronic structure of Si is the same as that of Ne. Recall that core electrons are generally considered inert with respect to determining reactivity and structure and, hence, we will ignore them. We therefore have four valence electrons. What is typically done is to consider these four electrons as distributed equally among the $3s$ and $3p$ orbitals. Thus, we can rewrite the electronic structure of a Si atom as $[Ne]3s^1 3p_x^1 3p_y^1 3p_z^1$.

From Ch. 11, the diamond structure has an fcc lattice with two atoms in the basis at $(0,0,0)$ and $(\frac{1}{4}, \frac{1}{4}, \frac{1}{4})$. The coordinates of the basis atoms are in units of the cubic unit cell edge. Therefore, with four atomic orbitals per atom and two atoms per unit cell, we have a total of eight atomic orbitals per unit cell. Already, then, we expect to find eight bands per unit cell. A conventional unit cell showing the lattice points and the basis atoms is reproduced in Fig. 13.1.

The tight-binding wavefunction is a linear combination of the atomic orbitals on each basis atom,

$$|\psi_k^{(n)}\rangle = \frac{1}{\sqrt{N}} \sum_{\text{cells}} \sum_{j=1}^{2} \sum_{\mu=s,p_x,p_y,p_z} c_{\mu,j}(k)\, e^{i\vec{k}\cdot\vec{r}_j} |\phi_\mu(j)\rangle \qquad (13.1)$$

where the outermost sum is over all unit cells in the crystal, the sum over j counts the two atoms in the basis, the sum over μ runs over the various types of atomic orbitals, $\vec{r}_j$ is a vector to the position of each atom in a unit cell, and N is the number of unit cells in a crystal. The superscript (n) is the band index, as discussed in Ch. 12. In order to make things simpler, we want the atomic orbitals to be orthonormal. This is generally not the case, but Löwdin† has showed that linear combinations of the original atomic orbitals can be used to create wavefunctions that are orthogonal to one another. We will keep our notation as above, remembering that the atomic basis functions are now functions with a slightly different mathematical form, but symmetry identical with that of the original atomic orbitals. Therefore,

$$\langle \phi_\mu(j)|\phi_\nu(m)\rangle = \delta_{jm}\delta_{\mu\nu} \qquad (13.2)$$

Just as we did for the simple two-band model in Ch. 12, we insert the wavefunction of Eq. 13.1 into the Schrödinger equation and take the overlap of it with each one of the atomic orbitals successively. This will generate 8 equations for the 8 atomic orbitals in a unit cell. If we only include on-site and nearest-neighbor interactions, then we will obtain an 8×8 secular determinant to solve for a nontrivial solution for the coefficients.

First consider the on-site matrix elements. Let us make the definitions $\alpha_{s,1} \equiv \langle \phi_s(1)|\mathcal{H}|\phi_s(1)\rangle$, $\alpha_{s,2} \equiv \langle \phi_s(2)|\mathcal{H}|\phi_s(2)\rangle$, $\alpha_{p_x,1} \equiv \langle \phi_{p_x}(1)|\mathcal{H}|\phi_{p_x}(1)\rangle$, etc. The on-site matrix elements between different orbitals will be zero for Si for symmetry reasons.

Exercise 13.1   Can you see why this must be true?

The nearest-neighbor hopping matrix elements will be somewhat more complicated in form than we are used to and we do not write them explicitly here. What's more, various authors have slightly different functional forms. Without explicitly writing down a functional form, however, we can understand schematically the types of interactions. These are shown in Fig. 13.2. The expressions in parentheses are those defined by Slater and Koster and they indicate the type of local bond formed.

Before we can properly discuss the band structure of Si, we need to

† P.-O. Löwdin, *J. Chem. Phys.* **18**, 365 (1950).

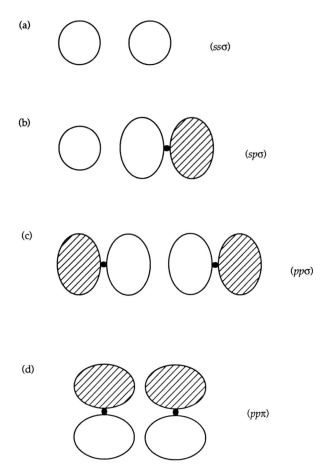

Fig. 13.2.   Schematic diagram of the four types of atomic orbital interactions possible in silicon.

construct the Brillouin zone. The real space primitive lattice vectors for the fcc lattice are:

$$\vec{a}_1^P = \frac{a}{2}(\hat{x} + \hat{y})$$
$$\vec{a}_2^P = \frac{a}{2}(\hat{y} + \hat{z})$$
$$\vec{a}_3^P = \frac{a}{2}(\hat{x} + \hat{z}) \qquad (13.3)$$

and are shown in Fig. 13.3. The reciprocal space lattice is found using Eqs. 8.7. The Brillouin zone is then constructed using the Wigner–Seitz method: sit on one lattice point and draw vectors to all surrounding lattice points.

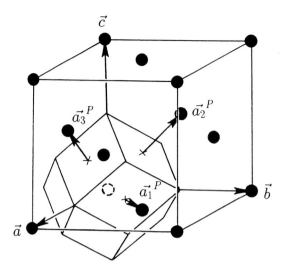

Fig. 13.3. Real space primitive lattice vectors for a face-centered cubic lattice, $\vec{a}_1^{\,P}, \vec{a}_2^{\,P}$, and $\vec{a}_3^{\,P}$. The conventional, multiply primitive lattice vectors are $\vec{a}, \vec{b}$, and $\vec{c}$.

Next, place a plane at the midpoint of these vectors and perpendicular to them. The smallest volume enclosed is the first Brillouin zone. This is shown in Fig. 13.4 with some of the special k points labeled. Notice that it does not matter what atoms constitute the basis, only the lattice is important for determining the Brillouin zone.

The 8×8 secular determinant looks schematically like Fig. 13.5, where we have blocked in each type of interaction. The exact determinant has geometrical terms for the placement of the two atoms in the unit cell. Chadi and Cohen† have examined this problem and broken the electronic structure calculation down into two parts. First of all, they ignored any s–p interaction and solved the band structure for each orbital type independently. They then included s–p interactions and solved the complete band structure problem.

In the first case, the 8×8 secular determinant factors into a 2×2 determinant and a 6×6 determinant for the s and p interactions, respectively. The band structure for the s part of the calculation is shown in Fig. 13.6 on the right, with the accompanying DOS on the left of Fig. 13.6. The p part of the calculation is shown in Fig. 13.7. The band labels are part of the BSW notation that we introduced previously. Decoupling the two types of orbital interactions is similar to projecting out just the s and p local DOS. Notice how the dominant low-energy contribution at the Γ point comes from the s

† D.J. Chadi and M.L. Cohen, *Phys. Stat. Sol. (B)* **68**, 405 (1975).

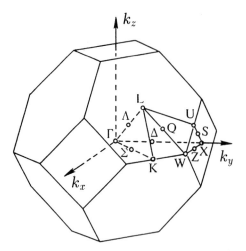

Fig. 13.4.   The first Brillouin zone for a face-centered cubic lattice. The special points are labeled.

	s_1	s_2	p_{x1}	p_{y1}	p_{z1}	p_{x2}	p_{y2}	p_{z2}
s_1	$\alpha_{s,1}-\mathcal{E}_k$	$(ss\sigma)$	0	0	0	⟵	$(sp\sigma)$	⟶
s_2	$(ss\sigma)$	$\alpha_{s,2}-\mathcal{E}_k$	⟵	$(sp\sigma)$	⟶	0	0	0
p_{x1}	0	↑	$\alpha_{x,1}-\mathcal{E}_k$	0	0	↑		
p_{y1}	0	$(sp\sigma)$	0	$\alpha_{y,1}-\mathcal{E}_k$	0	⟵	$(pp\sigma)$	⟶
p_{z1}	0	↓	0	0	$\alpha_{z,1}-\mathcal{E}_k$		$(pp\pi)$	↓
p_{x2}	↑	0	↑			$\alpha_{x,2}-\mathcal{E}_k$	0	0
p_{y2}	$(sp\sigma)$	0	⟵	$(pp\sigma)$	⟶	0	$\alpha_{y,2}-\mathcal{E}_k$	0
p_{z2}	↓	0	↓	$(pp\pi)$		0	0	$\alpha_{z,2}-\mathcal{E}_k$

Fig. 13.5.   Schematic of the 8 × 8 secular determinant for the band structure of crystalline silicon. Symbols are defined in the text and diagrams for the types of orbital interactions are shown in Fig. 13.2.

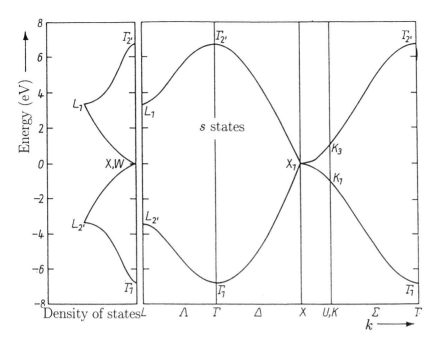

Fig. 13.6. Band structure for the *s* states of crystalline silicon from the calculation of Chadi and Cohen. The DOS is shown at left. From D.J. Chadi and M.L. Cohen, *Phys. Stat. Sol. (B)* **68**, 405 (1975). Used with permission.

states and that the top of the valence band comes from the *p* atomic orbitals; the *s* atomic orbitals are lower in energy than the *p* atomic orbitals.

For the full calculation, including *s*–*p* interaction, Chadi and Cohen actually included some second-neighbor interactions. The resulting band structure and DOS are shown in Fig. 13.8. Notice that only the valence bands are shown in this figure. Valence bands are considered to be the occupied bands, while conduction bands are the empty bands. Comparing Figs. 13.6, 13.7, and 13.8 makes several features clear. First of all, notice that the DOS goes to zero at the top of the valence band ($\Gamma_{25'}$ band). Although not shown here, there is an energy interval over which the DOS is zero. The band gap of Si at room temperature is approximately 1.2 eV, so that Si is a semiconductor. Secondly, there are three major peaks in the DOS: one marked $L_{2'}$, another marked L_1, and the top one marked X_4. These peaks are dominated by the *s*, *s*–*p*, and *p* parts of the atomic orbitals. When we study the DOS of amorphous Si later in this chapter, we will see remnants of these features. What is not apparent from Fig. 13.8 is that Si is an *indirect band gap* material; the highest-energy valence band is not at the same *k* point as the

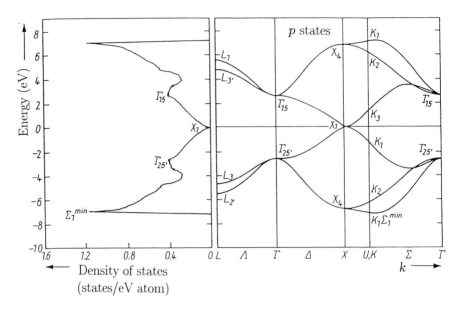

Fig. 13.7.   Band structure for the *p* states of crystalline silicon from the calculation of Chadi and Cohen. The DOS is shown at left. From D.J. Chadi and M.L. Cohen, *Phys. Stat. Sol. (B)* **68**, 405 (1975). Used with permission.

lowest-energy conduction band. The highest-energy valence band point is at the Γ point, while the lowest-energy conduction band point is found along the X direction. This is to be contrasted with the case of GaAs, which is a *direct band gap* material; the highest-energy valence band occurs at the same *k* point as the lowest-energy conduction band.

Exercise 13.2   Apply the moments theorem to Si in the diamond structure to see whether you can verify some of the features found in the total DOS.

13.3 Hydrogen chemisorption on Si(111) surface

By comparison with electronic structure calculations of bulk materials, calculations on surfaces are actually quite difficult. The main problem is that it is more cumbersome mathematically to deal with lack of periodicity in the direction perpendicular to the surface. In addition, one of the pitfalls that we must recognize is that surfaces are not simply passive receptors on which, for example, reactions occur. Surfaces can relax and reconstruct. In fact, one of the triumphs of the scanning tunneling microscope has been its elucidation of the complex reconstructions of semiconductor surfaces.

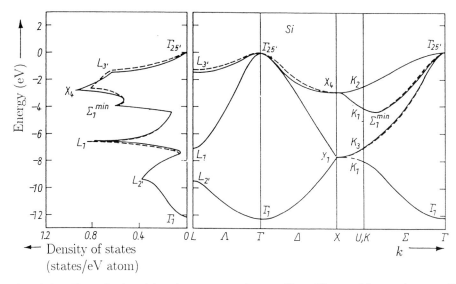

Fig. 13.8. The calculated band structure of crystalline silicon with *s* and *p* coupling. From D.J. Chadi and M.L. Cohen, *Phys. Stat. Sol. (B)* **68**, 405 (1975). Used with permission.

Why might we expect a surface to reconstruct? Consider the example of Si in the diamond structure, Fig. 13.1. Cleave the crystal at the bond midpoint in a plane perpendicular to the [111] direction. The exposed, unrelaxed, and unreconstructed surface is shown from the side in Fig. 13.9. We are left with a surface in which one orbital, populated by one electron, sticks out perpendicularly. This feature is called a *dangling bond*. Clearly this arrangement is unfavorable, for if a system can find any way at all, it will bond! Furthermore, we can imagine that this surface is extremely reactive. What we will examine in this section is the reaction of H with this idealized Si(111) surface following the tight-binding calculations of Pandey.†

Just as Miller indices are used to identify planes and directions in bulk crystals, there is a system for identifying surface structures. Why do we need a nomenclature for surfaces? Because often the reconstructed surface has different symmetry (usually lower) from that of the unreconstructed surface. We describe briefly the nomenclature developed by Wood. The surface structure is designated with respect to the bulk unit cell. If the surface shows no reconstruction, then it is called a (1 × 1) reconstruction. Most metal surfaces do not reconstruct. For example, the {110} and {211} surfaces of

† K.C. Pandey, *Phys. Rev. B* **14**, 1557 (1976).

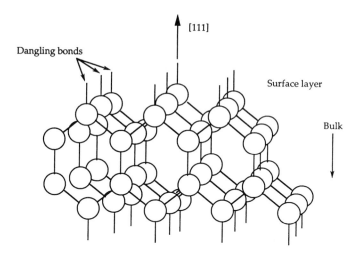

Fig. 13.9.   Schematic diagram of the side view of the (111) surface of diamond structure silicon. Dangling bonds are represented as incomplete bonds.

W do not reconstruct; the surface structure is thus designated $\{110\}(1 \times 1)$ or $\{211\}(1 \times 1)$.

For surfaces that do reconstruct, let us designate $\vec{a}_B$ and $\vec{b}_B$ as the bulk lattice vectors and $\vec{a}_S$ and $\vec{b}_S$ as the surface lattice vectors. Then if

$$\begin{aligned}\vec{a}_S &= m_a \vec{a}_B \\ \vec{b}_S &= m_b \vec{b}_B\end{aligned} \tag{13.4}$$

then the surface unit cell is $(m_a \times m_b)$.

For systems in which there is an adsorbed species, the same notation is used. H atoms, as we have seen, are a particularly good example of 'terminators' on our Si(111) surface. This surface is designated Si(111)(1 × 1)–H. For a surface unit cell that is not in complete registry with the underlying substrate, the nomenclature is only slightly more complicated. In that case, we define

$$\begin{aligned}\vec{a}_S &= m_{11}\vec{a}_B + m_{12}\vec{b}_B \\ \vec{b}_S &= m_{21}\vec{a}_B + m_{22}\vec{b}_B\end{aligned} \tag{13.5}$$

The subscripts on the factors m indicate that the surface unit cell vectors are not parallel to the bulk unit cell vectors. Just as there are centered unit cells in the bulk, there are also centered unit cells for surfaces. We will not go into further details here.

How do we set up a surface electronic structure calculation? We want to preserve the periodicity in the plane of the surface. To do so, we perform calculations on a slab of material so that we actually have *two* surfaces. These two surfaces have the structure in Fig. 13.9 and it is assumed that each dangling bond is saturated by a H atom. The surfaces are therefore covered with a monolayer of H. How thick should this slab be? That's a question to which we do not *a priori* know the answer. All we can say definitively is that the electronic structure at the center of the slab should resemble the electronic structure of bulk crystalline Si. In order to assure this requirement, the number of layers in the slab must be increased until the electronic structure at the center does not change.

As we discussed in the section on bulk crystalline Si, we need to know the Hamiltonian matrix elements. The assumption underlying tight-binding surface calculations is that matrix elements are transferable from one environment to another as long as the bond lengths remain unchanged. Matrix elements for bulk crystalline Si are therefore used in the surface calculation. The matrix elements are determined initially by fitting the band structure to other calculations or to experiment. Finally, the Si–H interaction matrix elements are found by fitting to Si–H bonds in molecular SiH_4. The Si–H bond length for the surface calculation is therefore set to the same value as in the molecule, 1.48 Å.

The surface bands for H chemisorbed on a Si(111) surface are shown in Fig. 13.10. This figure is slightly more complicated than the band structure of bulk Si. In the first place, the Brillouin zone of the surface is two-dimensional. Since the (111) surface is close-packed, the real space lattice is hexagonal and, hence, the reciprocal space lattice is hexagonal. This first Brillouin zone and special k points are shown in Fig. 13.11. In turn, reciprocal lattice vectors are two-dimensional and are denoted by $\vec{k}_{\parallel}$ to indicate that they are parallel to the surface. Surface states are distinguished from bulk states in that the former arise from splitting off of the bulk states due to a lowering of symmetry and the formation of a Si–H bond. From Fig. 13.10, there are surface bands only in regions for which there are gaps in the bulk states. These regions are white in Fig. 13.10. There are four surface bands found and they are shown by heavy lines: three at the K point and one in the vicinity of the J point.

The local densities of states for the H monolayer and three successive layers of Si (into the bulk) are plotted in Fig. 13.12. The local DOS of a bulk Si atom is superimposed on each figure. The H monolayer local DOS is characterized by a prominent peak at about −5 eV. This peak corresponds to the Si–H bond. Si–H bonding is also reflected in the second layer, which is the first layer of Si atoms. Here, too, the local DOS reflects a sharp

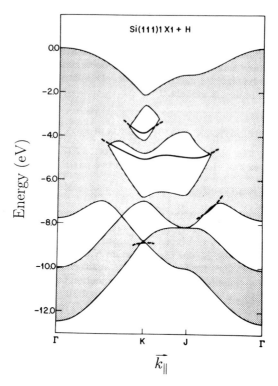

Fig. 13.10. Surface band structure for hydrogen adsorbed on a Si(111)(1 × 1) surface. $\vec{k}_\parallel$ is the wavevector parallel to the surface. From K.C. Pandey, *Phys. Rev. B* **14**, 1557 (1976). Used with permission.

peak at about −5 eV. Notice, however, that the local DOS has features which resemble those of the bulk. This resemblance becomes even more pronounced for the third and fourth layers. By the time we reach the fourth layer, the sharp peak at −5 eV, which is a signature of surface states, is greatly reduced in amplitude.

For all of these calculations only the valence bands, those bands that are occupied, are calculated. This is a reflection of our choice of basis orbitals. In particular, because we have included only *occupied* atomic orbitals in the basis, only occupied bands can be calculated with any accuracy. It is possible, however, to calculate the higher-lying unoccupied conduction bands within a tight-binding model by using an extended basis set.

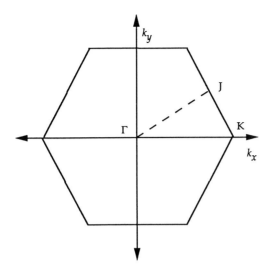

Fig. 13.11. Hexagonal Brillouin zone for the Si(111)(1 × 1) surface.

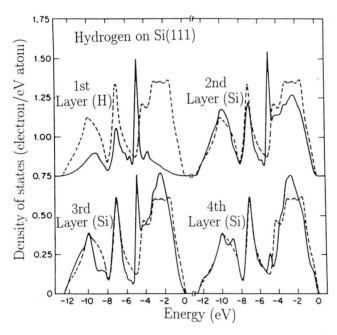

Fig. 13.12. Local density of states for the hydrogen monolayer and three successive silicon layers. Superimposed on each figure is the DOS of bulk crystalline silicon. From K.C. Pandey, *Phys. Rev. B* **14**, 1557 (1976). Used with permission.

Fig. 13.13.   Diagram showing the ring structure of a buckyball, C_{60}. A carbon atom sits at each vertex. Figure courtesy of S.J. Townsend.

13.4 Buckyballs and other novel forms of carbon

We shift attention now from extended structures back to molecules. But these molecules have a number of interesting properties and they have analogs in solids. Although only recently discovered, the class of pure carbon materials known as 'fullerenes' were rather fancifully proposed to exist in an article first published in *New Scientist* in 1966.†

The most common fullerene is a truncated icosahedral structure, shown in Fig. 13.13. It contains 60 carbon atoms arranged in a cage-like structure that consists of 20 six-membered rings and 12 five-membered rings. All of the atoms have identical surroundings. There are clear comparisons between features of this structure and benzene or graphite. The name 'fullerene' is taken from Buckminster Fuller, who studied icosahedral structures as housing alternatives in the 1970s. The C_{60} molecule is commonly called a 'buckyball'.

The simple tight-binding calculations that we have used can be applied to study this system as well. Because each atom is at the vertex of three rings and the electronic structure of C is $[He]2s^2 2p^2$, three valence electrons are used in σ bonding to neighbors. The fourth valence electron can be considered to occupy the $2p_z$ atomic orbital, where we have chosen the z axis on *each* atom to be perpendicular to the surface of the cage; these electrons will be involved in π bonding over the surface of the cage. The

† *New Scientist*, 3 November 1966.

π bond tight-binding wavefunction is easily written down as a sum over p_z atomic orbitals on each of the 60 carbon atoms,

$$|\psi\rangle = \frac{1}{\sqrt{60}} \sum_{j=1}^{60} c_j |p_z(j)\rangle \qquad (13.6)$$

Because this system is molecular and does not have translational symmetry, we cannot use the concept of reciprocal space. Nonetheless, $|\psi\rangle$ can be inserted into the Schrödinger equation, and, making our usual assumptions of neglect of overlap and nearest-neighbor interactions only, the secular determinant can be set up. The cumbersome part of this calculation is that the secular determinant is now 60×60. Using any one of the symbolic manipulation programs available on many computers now, it is a simple matter to solve for the 60 roots of the determinant.

A schematic diagram for the energy levels so obtained is shown in Fig. 13.14. The character table of the icosahedral group is given in Table 13.1. Each of the states in Fig. 13.14 is labeled by the appropriate irreducible representation label and the degeneracies can be determined either from these labels or from the number of states for each level in Fig. 13.14.

All energies in Fig. 13.14 are in units of the unknown hopping matrix element, β. By the Pauli exclusion principle, each state is populated by two electrons for a total of 60 electrons. It is interesting to note that there is a gap between filled and empty states in C_{60}. The ground state of C_{60} is therefore predicted to be insulator-like and the molecule should be *diamagnetic*. That is, all spins are paired in the ground state and, hence, buckyballs have no net magnetic moment.

The wavefunctions for this system are not so easily visualized because of the spherical geometry of the molecule. However, examination of the eigenvectors generated reveals that the lowest-energy state, which is nondegenerate, has all p_z atomic orbitals in phase with the same amplitude, so that this represents a purely bonding state. Other states in the valence band are different combinations which decrease in their bonding character as we increase in energy. The highest unoccupied state shifts phases alternately and so is completely antibonding.

13.5 Amorphous silicon

As the last case study in this chapter we consider what happens when we lose translational symmetry completely. As a specific example, we examine amorphous silicon. Amorphous materials present an intriguing challenge for electronic structure calculations. As we noted in Ch. 1, amorphous

Table 13.1. *Character table for the icosahedral group*

I_h	E	$12C_5$	$12C_5^2$	$20C_3$	$15C_2$	i	$12S_{10}^3$	$12S_{10}$	$20S_6$	15σ	$\alpha = \frac{1}{2}(1+\sqrt{5})$	$\beta = \frac{1}{2}(1-\sqrt{5})$
A_g	1	1	1	1	1	1	1	1	1	1		$x^2+y^2+z^2$
T_{1g}	3	α	β	0	-1	3	β	α	0	-1	(R_x, R_y, R_z)	
T_{2g}	3	β	α	0	-1	3	α	β	0	-1		
G_g	4	-1	-1	1	0	4	-1	-1	1	0		
H_g	5	0	0	-1	1	5	0	0	-1	1		$\left.\begin{array}{l} 2z^2-x^2-y^2 \\ x^2-y^2 \\ xy \\ yz \\ xz \end{array}\right\}$
A_u	1	1	1	1	1	-1	-1	-1	-1	-1		
T_{1u}	3	α	β	0	-1	-3	$-\beta$	$-\alpha$	-1	1	(x, y, z)	
T_{2u}	3	β	α	0	-1	-3	$-\alpha$	$-\beta$	0	1		
G_u	4	-1	-1	1	0	-4	1	1	-1	0		
H_u	5	0	0	-1	1	-5	0	0	1	-1		

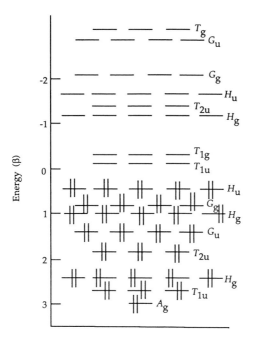

Fig. 13.14.   Schematic energy level diagram showing the allowed states for C_{60}.
Energies are in units of the hopping matrix element, β. Labels are assigned to each
of the states from the irreducible representations of the icosahedral group. Occupied
states are indicated by two vertical lines to represent two electrons.

materials are characterized by local order, so that each Si atom sits in a
nearly tetrahedral local environment, but they lack long-range order. The
major consequence is that we cannot apply Bloch's theorem to find the
electronic structure of the material. Hence, the ideas of Brillouin zones and
band structures are meaningless. Nonetheless, we show in this section how
we can use the tight-binding method to write down a wavefunction and how
the DOS gives us an indication of the electronic properties of disordered
materials.

But what exactly is the 'structure' of an amorphous material? Because of
the disorder inherent in such materials, any description of structure must
be statistical in nature. One quantity that is frequently used is the *radial
distribution function* (RDF). If we were to imagine sitting on one atom
inside a sample of amorphous Si, then the RDF gives us the probability of
locating neighboring atoms. The RDF obtained from electron diffraction
experiments on crystalline and amorphous diamond is shown in Fig. 13.15.
The dashed line, for crystalline diamond, shows finite probability at well-

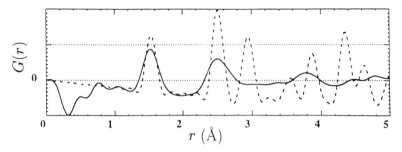

Fig. 13.15. The radial distribution function (RDF) for crystalline (dashed line) and amorphous (solid line) carbon. From D.A. Muller, Master's Thesis, University of Sydney.

defined separations, indicating an ordered structure. The solid line, by contrast, retains the first two peaks, but loses structure for larger separations. This indicates that the first and second neighbors in the amorphous sample are at approximately the same separations as in the crystal, but that any longer-range order is absent.

In 1932, Zachariasen† introduced the concept of a continuous random network (CRN) to model the structure of inorganic glasses. A CRN model of amorphous Si is shown in Fig. 13.16. This model assumes that each atom is bonded to four nearest neighbors. Disorder is introduced through variations in the bond angles; bond lengths may be kept at their crystalline Si values or they may be varied as well. Periodic boundary conditions are imposed on the models to eliminate free surfaces. It must be true that this 'supercell' is large enough so that the underlying periodicity does not affect calculated properties. As with the thickness of the slab for surface calculations, cells of increasingly larger sizes are used until the results converge.

Given this model structure, a tight-binding wavefunction for the entire supercell can be written down and inserted into the Schrödinger equation, and the secular determinant solved for the eigenvalues and eigenvectors. One problem that must be solved here is similar to that encountered in the surface problem. Namely, we must figure out what the Hamiltonian matrix elements are. If we include only first-nearest-neighbor interactions, then bond angle variations from the perfect tetrahedral angle are not considered. Thus, only bond length deviations from the ideal 2.35 Å need to be taken into account. What has often been done follows from an observation made by Harrison.‡ He noticed that the matrix elements varied approximately with the inverse

† W.H. Zachariasen, *J. Am. Chem. Soc.* **54**, 3841 (1932).
‡ S. Froyen and W.A. Harrison, *Phys. Rev. B* **20**, 2420 (1978).

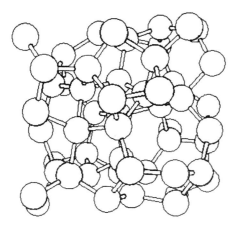

Fig. 13.16.   A computer-generated CRN model of amorphous silicon. Periodic boundary conditions are applied at the surfaces of the cell.

square of the atomic separation. Therefore, what is most commonly done is to take the matrix elements found for crystalline Si and scale them for each bond length within the CRN model.

The energy eigenvalues are most conveniently displayed as a total DOS. We show a schematic for an amorphous silicon model in Fig. 13.17(b). Also included in this figure is the calculated DOS for crystalline Si. Notice that the sharp peaks present in the crystalline Si DOS are absent in the amorphous Si DOS. What is intriguing, however, is that amorphous Si still contains a band gap. Recall from our discussion of the two-band model that a band gap arises because of local chemical interaction and not because of long-range order. We would therefore expect amorphous Si to possess a band gap, since the local order remains very similar to that in crystalline Si.

If we examine the wavefunctions obtained from solving the secular determinant, then we find that the bottom of the valence band is dominated by contributions from the Si s atomic orbitals and the top of the valence band is dominated by contributions from the Si p atomic orbitals. Again, this is similar to what is found for crystalline Si. We find an interesting difference from crystalline Si, however. As we discussed in Ch. 10, the conductivity of a perfect crystal at 0 K is infinite. In turn, this means that the bands allow the electrons to be delocalized over the entire crystal. On the other hand, amorphous Si contains many defects. These defects are manifest in the electronic structure of the material by the presence of another kind of 'gap', a so-called *mobility gap*. The complete DOS for crystalline Si and amorphous Si are compared in Fig. 13.18. At the band edge in crystalline Si

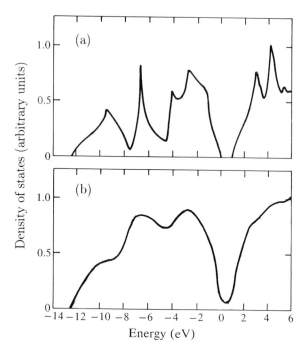

Fig. 13.17. Density of states for crystalline silicon (a) and amorphous silicon (b).

are the van Hove singularities first introduced in Ch. 10. However, because of the disorder in amorphous Si, sharp band edges do not exist and are, instead, replaced by a mobility edge. The gap between mobility edges is the mobility gap. The fundamental nature of the wavefunctions changes as we move through the mobility edge. Bands below the mobility edge (for the valence band) are extended over some distance in the material. Bands above the mobility edge are *localized* in the vicinity of just a few atoms. Current is conducted very differently in these two types of states. The extended states carry current just as in the case of a crystal. But electrons must hop from one localized state to another one. The conductivity of an amorphous material thus has a very different functional dependence on temperature from that of a crystal.

Problems

13.1 Use the moments theorem to find the moments of the DOS for crystalline Si up to the fourth moment ($\mu^{(4)}$). How do your results compare with the DOS found from application of the tight-binding

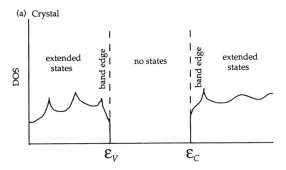

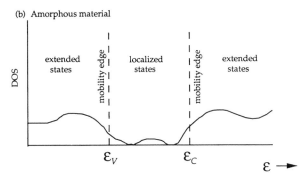

Fig. 13.18. Schematic diagram showing the difference between features in the density of states of (a) crystalline silicon and (b) amorphous silicon. The latter lacks sharp band edges; instead they are replaced by mobility edges. $\mathscr{E}_C$ ($\mathscr{E}_V$) is the conduction (valence) band edge.

method? Now use the moments theorem to find the moments of the DOS for a continuous random network model of amorphous Si. Can you distinguish between the moments of the DOS for crystalline Si and amorphous Si? What do you need in order to be able to do so?

13.2 GaAs is a typical III–V compound semiconductor. It has the zinc sulfide crystal structure; that is, an fcc lattice with a two-atom basis. (See Fig. 11.9.)

(a) Following the calculations outlined in the text for crystalline Si, write down the tight-binding wavefunction for GaAs. What are the valence electrons? What are the core electrons that you can ignore?

(b) Take the overlap of your wavefunction with each of the atomic orbitals successively. How many equations do you generate for the atomic orbitals in a single unit cell? Include only on-site and nearest-neighbor interactions. Define the on-site matrix elements in the usual way and draw schematic diagrams for the nearest-neighbor hopping integrals.

(c) What is the Brillouin zone for GaAs?

(d) Can you factor the secular determinant into two contributions, as was done in the case of Si? If so, *sketch* the DOS for the s and p contributions separately and then for the interacting problem. You are not required to calculate the band structure explicitly. Which atom (*i.e.*, Ga or As) contributes to the bottom of the valence band? To the top of the valence band?

13.3   Consider an AB crystal with the NaCl structure. An s atomic orbital of energy $\mathcal{E}_A$ is associated with each A atom and an s atomic orbital of energy $\mathcal{E}_B$ is associated with each B atom. Assume that $\mathcal{E}_A > \mathcal{E}_B$. Only the hopping matrix elements between nearest neighbors are nonzero and they are set equal to β ($\beta < 0$). You may also assume that the atomic basis set is orthonormal.

The wavefunctions of the AB crystal can be written in the form:

$$|\psi_k^{(n)}\rangle = \frac{1}{\sqrt{N}} \sum_m (c_A^{(n)}(\vec{k})|\vec{R}_m, A\rangle + c_B^{(n)}(\vec{k})|\vec{R}_m + \vec{\tau}, B\rangle) e^{i\vec{k}\cdot\vec{R}_m} \quad (P13.1)$$

where n is the band index, N is the total number of lattice sites, $\vec{R}_m$ is the position vector of the mth lattice site, and $\vec{\tau}$ is the position of the B atom relative to the A atom in each primitive unit cell. For our case, $\vec{\tau}$ is $(a/2)[100]$ where a is the lattice parameter.

(a) Using the Schrödinger equation, find the two secular equations.

(b) Find the two roots for the energy. What are the energy ranges for the two bands? What is the value of the energy gap?

(c) Sketch *qualitatively* the total density of states.

13.4   Consider a perfect Si crystal. Cut the crystal so that you expose a (001) plane.

(a) Draw a schematic diagram of the side view of the surface, assuming that the surface does not relax or reconstruct. How many dangling bonds per surface atom are there?

(b) Since the surface as prepared in part (a) contains dangling bonds, we will expect it to reconstruct. How might you expect it to do so? Draw schematic diagrams indicating possible scenarios.

(c) Consult D.J. Chadi and M.L. Cohen, *Phys. Stat. Sol. (B)* **68**, 405 (1975). Describe these electronic structure calculations.

13.5   In addition to the buckyball and other structures similar to it, carbon 'needles' have also been discovered. These needles consist of hollow, rolled sheets of graphite, only a few nanometers in diameter. Some of the electronic properties of graphitic needles have been investigated in the paper by Hamada *et al.*, *Phys. Rev. Lett.* **68**, 1579 (1992). Following this article, construct a model of a graphitic tubule. Clearly indicate the meaning of the indices introduced in the article. Using the arguments laid out in the article and the results of our own discussion of bonding in a graphite sheet (Problem 10.5), discuss the band structure of the $B(1,0)12$ and $B(1,0)13$ cases, especially with regard to the electrical conductivity of the needles. What types of tubules are predicted to be metals? Why?

13.6 The remarkable stability of some cyclic hydrocarbons has been known to chemists for quite some time. It is believed that the ability of the electrons involved in π bonding in these systems to delocalize over many atoms is the source of this stability. Referring to the article by Jun-ichi Aihara, 'Why aromatic compounds are stable' in the March 1992 issue of *Scientific American*, discuss Aihara's definition of the delocalization energy. In particular, set up the 'characteristic polynomial' for cyclopropyl and find its roots. Next set up the 'reference polynomial' for the 'quasichain' molecule with the same number of carbon atoms and find its roots. (See Aihara's article, p. 68.) By comparing these two numbers, you will find a value for the topological resonance energy. This number has been used to explain the stability of a wide range of molecules, including the fullerenes. Can you indicate how to set up the problem for C_{60}?

14

Free-electron and nearly-free-electron systems

14.1 Introduction

We discussed the conductivity of materials using band-filling arguments in Ch. 12. There we argued that at 0 K and in the absence of defects, a partially filled band has infinite conductivity. In addition, an insulator will always have a completely filled band. However, as we pointed out, the converse is not necessarily true. Elements on the left-hand side of the periodic table, in Groups 1A, 2A, and 3A, have valence electron configurations of the form [noble gas]$ns^l np^m$, where n is the principal quantum number and l and m are integers such that $l = 1, 2$, $m = 0, 1$. These elements thus have a deficit of valence electrons compared with orbitals available.

We might be tempted to conclude that Group 2A elements such as Be and Mg are insulators because the s atomic orbitals give rise to a filled band. However, the overlap between s and p atomic orbitals of neighboring atoms is really quite large and this interaction yields widely dispersed bands of mixed s and p character. With so few valence electrons and so many bands available, even Group 2A elements are good metals. As an additional consequence, the electrons in these elements do not retain much of their parent atoms' identity. We have stressed the tight-binding approach to understanding electronic structure. However, it is clear that atomic orbitals are not an appropriate set of basis functions for metals. A better basis set consists of functions that delocalize the electrons more widely throughout the crystal. Precisely what these functions are is one of the questions to be investigated in this chapter.

The small number of valence electrons relative to the number of bands available is also reflected in the structures adopted by metals. It is most advantageous to share electrons amongst as many neighbors as possible. Metals therefore tend to adopt close-packed structures as opposed to the more open

226

and lower coordinated diamond and zinc sulfide structures preferred by C, Si, Ge, and GaAs, for example.

Drude put forward the first model of electronic structure of metals in 1900. His theory treated the free electrons as a 'gas' and he applied Maxwell's then recently developed kinetic theory of gases to determine properties of this gas. Drude's model failed in a number of troubling ways, however, particularly in predicting thermal properties of metals. It was not until the discovery of quantum theory and the Pauli exclusion principle that these anomalies were rectified. Sommerfeld recognized that the Maxwell–Boltzmann distribution of electron velocities should be replaced by the Fermi–Dirac distribution.

In this chapter we will discuss Sommerfeld's theory of the electron gas for free-electron metals. We will also discuss how to modify this theory for metals with 'nearly-free electrons', *i.e.*, those that do feel the periodic potential of the atoms. In Ch. 15 we will discuss the transition metals, for which the free-electron and nearly-free-electron models show the most serious deficiencies.

14.2 Free-electron gas

We continue to consider electrons as independent, noninteracting particles, but recognize that quantum mechanics requires the occupancy of levels at finite temperatures to be governed by the Fermi–Dirac distribution function,

$$f(\mathscr{E}) = \frac{1}{1 + \exp[(\mathscr{E} - \mathscr{E}_{\mathrm{F}})/k_{\mathrm{B}}T]} \qquad (14.1)$$

where $\mathscr{E}_{\mathrm{F}}$ is the Fermi energy, k_{B} is Boltzmann's constant, and T is the absolute temperature. The Fermi–Dirac distribution function is plotted in Fig. 14.1. The line labeled T_1 is valid for $T \approx 0$ K and it shows that states close up to the Fermi energy are occupied and a small fraction of states above $\mathscr{E}_{\mathrm{F}}$ are also occupied. All other states higher in energy are empty. At $T_2 > 0$ K, the sharp edges of the Fermi function get rounded off; electrons can be excited from states below $\mathscr{E}_{\mathrm{F}}$ to states above $\mathscr{E}_{\mathrm{F}}$. At even higher temperatures, $T_3 > T_2$, occupation of states becomes even more smeared out in the vicinity of the Fermi energy. The explicit functional form in Eq. 14.1 can be 'derived' on the basis of statistical and thermodynamics arguments, but we will not pursue those here.

Let us turn to a consideration of the properties of the free-electron gas. We will examine the ground state properties (*i.e.*, $T=0$ K) first before investigating finite temperatures. Imagine that our metallic system consists of a collection of noninteracting electrons in the *constant* background potential

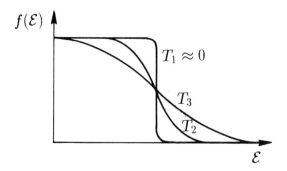

Fig. 14.1.   The Fermi–Dirac distribution function. The slightly rounded step function form is valid at $T_1 \approx 0$ K, while the smoother curves (T_2, T_3) are valid for $T > 0$ K.

of nuclei. We seek the electron wavefunctions and eigenvalues for this system. We can write the time-independent Schrödinger equation as

$$\left(-\frac{\hbar^2}{2m}\nabla^2 + V\right)\psi(\vec{r}\,) = \mathscr{E}\psi(\vec{r}\,) \tag{14.2}$$

where $\psi(\vec{r}\,)$ is the one-electron wavefunction and V is the constant atomic potential. Remember that the atomic potential is constant, and not periodic, because, within this model, the electrons do not feel the atoms. We can simplify Eq. 14.2 by recognizing that a constant potential just shifts the zero of energy. We can therefore set our zero immediately to V and thus eliminate this term from Eq. 14.2.

Next, we must find appropriate boundary conditions for Eq. 14.2. Just as we did for the simple tight-binding model of H $1s$ orbitals, we can apply periodic boundary conditions. In the present case, we can confine the electron gas to a *cube* of side L. The potential inside the cube is everywhere constant (taken here as zero), while it rises to infinity at the walls. In this context, periodic boundary conditions are often referred to as the *Born–von Kármán boundary conditions*. The wavefunction thus has translational symmetry with period L,

$$
\begin{aligned}
\psi(x+L, y, z) &= \psi(x, y, z) \\
\psi(x, y+L, z) &= \psi(x, y, z) \\
\psi(x, y, z+L) &= \psi(x, y, z)
\end{aligned}
\tag{14.3}
$$

A general, normalized solution of Eq. 14.2 is

$$\psi_{\vec{k}}(\vec{r}\,) = \frac{1}{\sqrt{V}}e^{i\vec{k}\cdot\vec{r}} \tag{14.4}$$

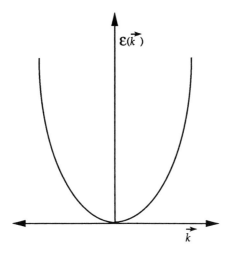

Fig. 14.2. Functional dependence of the energy $\mathscr{E}$ on wavevector for a free-electron gas.

where $V = L^3$ and $\vec{k}$ is just the wavevector that we introduced previously in Ch. 8. The eigenvalues are given by

$$\mathscr{E}(\vec{k}) = \frac{\hbar^2 k^2}{2m} \qquad (14.5)$$

Exercise 14.1   Verify Eq. 14.5 and that Eq. 14.4 is a solution to the Schrödinger equation.

The energy eigenvalues for the free-electron gas are plotted in Fig. 14.2. Note that for the wavefunction of Eq. 14.4, the wavevector *is* an eigenvalue of the momentum operator, $\vec{p} = (\hbar/\imath)\nabla$, since

$$\frac{\hbar}{\imath}\nabla e^{\imath\vec{k}\cdot\vec{r}} = \hbar\vec{k} e^{\imath\vec{k}\cdot\vec{r}} \qquad (14.6)$$

so that the eigenvalue is $\hbar\vec{k}$. Unlike the Bloch wavefunction, the free-electron wavefunction does have a well-defined momentum, $\vec{p} = \hbar\vec{k}$, which is in accord with our classical interpretation of momentum.

If we now apply the boundary conditions of Eq. 14.3, then $\vec{k}$ is limited to discrete values,

$$\psi_{\vec{k}}(x, y, z) = \frac{1}{\sqrt{V}} e^{\imath(k_x x + k_y y + k_z z)} = \frac{1}{\sqrt{V}} e^{\imath(k_x(x+L) + k_y y + k_z z)} \qquad (14.7)$$

with similar equations in y and z. Therefore,

$$e^{\imath k_x L} = e^{\imath k_y L} = e^{\imath k_z L} = 1 \qquad (14.8)$$

or

$$k_x = \frac{2\pi}{L}n_x$$

$$k_y = \frac{2\pi}{L}n_y$$

$$k_z = \frac{2\pi}{L}n_z \tag{14.9}$$

where n_x, n_y, and n_z are integers.

Exercise 14.2   Sketch the form of the wavefunction for selected values of the integers n_x, n_y, and n_z. What features can you discern as the integers increase in magnitude?

Points in a three-dimensional cubic reciprocal space represent each free-electron energy in precisely the same way as for tight-binding systems. The energy of any electron is, from Eqs. 14.5 and 14.9,

$$\mathcal{E}(\vec{k}) = \frac{\hbar^2}{2m}\left(\frac{2\pi}{L}\right)^2 (n_x^2 + n_y^2 + n_z^2) \tag{14.10}$$

A three-dimensional square array of reciprocal lattice points is shown in the top of Fig. 14.3. In two dimensions, a slice through this three-dimensional array forms a square lattice. See Fig. 14.3, bottom. From Eq. 14.10, the energy of any electron depends only on its *distance* from the origin in reciprocal space. Furthermore, if we occupy levels following the Pauli exclusion principle, then the Fermi surface is a *sphere*. The sphere at the Fermi energy, $\mathcal{E}_F$, is of radius k_F, the Fermi wavevector.

The number of electrons, N, that can be accommodated inside the Fermi sphere is simply twice the number of reciprocal lattice points inside the sphere of radius k_F. The number of reciprocal lattice points is given by the volume of the Fermi sphere divided by the volume of k space per lattice point, which is $(2\pi/L)^3$. Thus,

$$N = 2\left(\frac{4\pi k_F^3}{3}\right)\left(\frac{L}{2\pi}\right)^3$$

$$= \frac{8\pi}{3}\left(\frac{2m\mathcal{E}_F}{h^2}\right)^{\frac{3}{2}} L^3 \tag{14.11}$$

If we write the electron density as $\rho = N/L^3$, then

$$\mathcal{E}_F = \frac{h^2}{2m}\left(\frac{3\rho}{8\pi}\right)^{\frac{2}{3}} \tag{14.12}$$

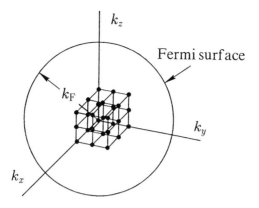

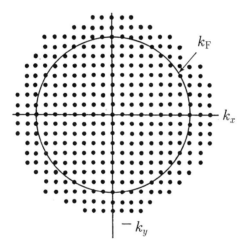

Fig. 14.3.   Schematic diagram of the cubic three-dimensional array of allowed energies in reciprocal space (top). Schematic diagram of a two-dimensional slice through the three-dimensional array (bottom).

The Fermi energy at absolute zero therefore depends only on the electron density, which is directly related to the number of valence electrons.

We can also find the density of states for the free-electron model, since the density of states is just the number of states per unit volume available between energy $\mathscr{E}$ and $\mathscr{E} + d\mathscr{E}$. From previous discussion, however, we know that the total number of electrons is just the integral over the density of states from zero up to the Fermi energy. Taking the first derivative of Eq. 14.10,

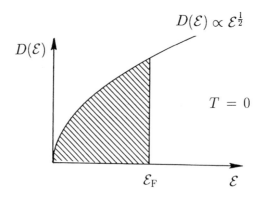

Fig. 14.4. Density of states for the three-dimensional free-electron gas.

then,

$$D(\mathcal{E}) = \frac{dN}{d\mathcal{E}}$$

$$= 4\pi \left(\frac{2m}{h^2}\right)^{\frac{3}{2}} \mathcal{E}^{\frac{1}{2}} \tag{14.13}$$

which is plotted in Fig. 14.4.

Exercise 14.3 Repeat the above derivation in one and two dimensions and plot the results for the density of states.

The band energy, which was introduced in Ch. 9, is just the sum over all occupied levels times the energy,

$$\mathcal{E}_{\text{band}} = \int_0^{\mathcal{E}_F} D(\mathcal{E}) \, \mathcal{E} \, d\mathcal{E} \tag{14.14}$$

Using the functional form for the DOS of a three-dimensional free-electron system (Eq. 14.13),

$$\mathcal{E}_{\text{band}} = \int_0^{\mathcal{E}_F} 4\pi \left(\frac{2m}{h^2}\right)^{\frac{3}{2}} \mathcal{E}^{\frac{3}{2}} \, d\mathcal{E}$$

$$= 4\pi \left(\frac{2\pi}{h^2}\right)^{\frac{3}{2}} \frac{2}{5} \mathcal{E}_F^{\frac{5}{2}} \tag{14.15}$$

The energy density per electron, $\mathcal{E}_{\text{band}}/\rho$, is thus

$$\frac{\mathcal{E}_{\text{band}}}{\rho} = \frac{3}{5} \mathcal{E}_F \tag{14.16}$$

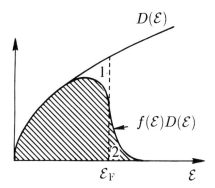

Fig. 14.5. The occupancy of states in the free-electron model at $T > 0$ K.

or, defining a Fermi temperature, $T_F \equiv \mathscr{E}_F/k_B$, we can rewrite Eq. 14.16 as

$$\frac{\mathscr{E}_{\text{band}}}{\rho} = \frac{3}{5}k_B T_F \tag{14.17}$$

which is a very different result from the energy per particle in a Maxwellian classical gas, $\frac{3}{2}k_B T$.

Exercise 14.4   Find an expression for the Fermi energy in terms of quantities you know, *i.e.*, fundamental constants, *etc*. For several free-electron metals, find the value of the Fermi energy and the Fermi temperature. How can you interpret the magnitude of the Fermi temperature that you obtain (it is large!)?

Suppose now that $T > 0$ K. What are the finite temperature properties of our electron gas? First, and most importantly, from Fig. 14.1, we can see that the Fermi–Dirac distribution changes only in the region right around the Fermi energy. The free-electron density of states at finite temperature is therefore altered from Fig. 14.4. The dotted line in Fig. 14.5 represents the occupancy of levels at $T > 0$ K. Note that this figure is *not* the density of states as some texts claim; it is the product of the density of states and the Fermi–Dirac distribution function.

Exercise 14.5   What is $k_B T$ at room temperature? Using your estimate above for $\mathscr{E}_F$, calculate specifically how much the Fermi–Dirac distribution function differs at room temperature from the $T = 0$ K distribution. Also, calculate the difference in energy between successive levels. How does this number compare with $k_B T$ at room temperature?

The most important property that is altered by using the (correct) Fermi–Dirac distribution function, as opposed to the Maxwell–Boltzmann distribution, is the electronic contribution to the constant-volume specific heat,

$$c_v = \left(\frac{\partial u}{\partial T}\right)_V \tag{14.18}$$

where u is the energy density, U/V, and the total internal energy of the system is just the sum over occupied levels, or the band energy. Note, however, that the equivalence between the internal energy and the band energy is only true within the independent electron approximation. For $T > 0$ K, we must amend Eq. 14.16 to reflect the change in occupation of the levels, so that

$$\mathscr{E}_{\text{band}} \equiv u = \int_0^{\infty} D(\mathscr{E}) f(\mathscr{E}) \, \mathscr{E} \, d\mathscr{E} \tag{14.19}$$

Integrals involving the Fermi distribution function can be evaluated with the aid of the Sommerfeld expansion (see Appendix 7). Using the results from Appendix 7, we can write

$$\mathscr{E}_{\text{band}} = N\mathscr{E}_{\text{F}}(T = 0) \left[1 + \frac{5\pi^2}{12}\left(\frac{k_B T}{\mathscr{E}_F}\right)^2\right] \tag{14.20}$$

It can be shown from more detailed analysis that Eq. 14.20 is correct to order T^2. $\mathscr{E}_F$ is defined as the Fermi energy at $T = 0$ K. The specific heat follows from Eq. 14.20 by taking the temperature derivative,

$$c_v = N\mathscr{E}_{\text{F}}(T = 0)\left(\frac{5\pi^2}{6}\right)\left(\frac{k_B}{\mathscr{E}_F}\right)^2 T \tag{14.21}$$

Using Eq. 14.17, we can rewrite Eq. 14.21 as

$$c_v = \frac{\pi^2}{2}k_B N \frac{T}{T_F} \tag{14.22}$$

so that the electronic contribution to the specific heat depends linearly on temperature. This result is smaller than the classical Maxwellian gas prediction by a factor proportional to T/T_F. The linear temperature dependence has been tested and verified for a number of simple metals. Notice that measurement of the low-temperature specific heat (so that ionic contributions are minimized) yields a coefficient that is proportional, through Eq. 14.13, to the density of states at the Fermi energy.

14.3 Nearly-free electrons

What happens now if, for example, the ionic potential is not constant, nor is it so strong as to warrant use of the tight-binding approximation? We still assume that the potential is periodic. The kind of wavefunction that we imagine for such a system is a plane wave to first order, but modulated to reflect the periodic potential. As a starting point we thus take the wavefunction as a *linear combination of plane waves,*

$$\psi_{\vec{K}}(\vec{r}) = \sum_{\vec{k}} c_{\vec{k}} |\vec{k}\,\rangle \tag{14.23}$$

where $|\vec{k}\,\rangle = (1/V)e^{i\vec{k}\cdot\vec{r}}$. Inserting Eq. 14.23 into the Schrödinger equation and setting up the secular determinant, we find

$$\det(\mathcal{H}_{\vec{k}\vec{k}'} - \mathcal{E} S_{\vec{k}\vec{k}'}) = 0 \tag{14.24}$$

is the condition for a nontrivial set of coefficients. Here $\mathcal{H}_{\vec{k}\vec{k}'} \equiv \langle \vec{k} | \mathcal{H} | \vec{k}' \rangle$ and $S_{\vec{k}\vec{k}'} = \langle \vec{k} | \vec{k}' \rangle$, just as we have defined in earlier chapters.

Because our system still has translational symmetry, the wavefunctions $|\vec{k}\rangle$ must generate irreducible representations of the translation group (Ch. 8). What's more, from the theorems at the end of Ch. 8, the only way in which the matrix elements, $\mathcal{H}_{\vec{k}\vec{k}'}$,[†] and overlap integrals, $S_{\vec{k}\vec{k}'}$, can be nonzero is if $|\vec{k}\rangle$ and $|\vec{k}'\rangle$ *both* generate the *same* irreducible representation. When will this happen? From Ch. 8, $\vec{k}$ and $\vec{k}'$ generate the same irreducible representation when they differ by a reciprocal lattice vector,

$$\vec{K} = n_1 \vec{b}_1 + n_2 \vec{b}_2 + n_3 \vec{b}_3 \tag{14.25}$$

where n_1, n_2, and n_3 are integers and $\vec{b}_1, \vec{b}_2$, and $\vec{b}_3$ are reciprocal lattice vectors as defined by Eqs. 8.7.

Exercise 14.6   Show explicitly that if $\vec{k}' = \vec{k} + \vec{K}$, then $|\vec{k}'\rangle$ and $|\vec{k}\rangle$ generate the same irreducible representations. Hint: Showing this really comes down to showing that the translation operator operating on $|\vec{k}\rangle$ and $|\vec{k}'\rangle$ yields the same function.

The requirement that

$$\vec{k}' = \vec{k} + \vec{K} \tag{14.26}$$

means that *all* wavevectors in Eq. 14.23 are equivalent to a single $\vec{k}$. In turn, it means that all wavefunctions can be written as a sum of plane waves with

---

[†] Recall that $\mathcal{H}$ transforms as Γ_1, the totally symmetric irreducible representation.

a fixed wavevector $\vec{k}$ and summed over *reciprocal lattice vectors*. Thus, it is more correct to write

$$\psi_{\vec{k}}(\vec{r}) = \sum_{\vec{K}} c_{\vec{K}} |\vec{k} + \vec{K}\rangle \tag{14.27}$$

Hence, the secular determinant can be written as

$$\det(\mathscr{H}_{\vec{K}\vec{K}'} - \mathscr{E}S_{\vec{K}\vec{K}'}) = 0 \tag{14.28}$$

where

$$\mathscr{H}_{\vec{K}\vec{K}'} \equiv \langle \vec{k} + \vec{K} | \mathscr{H} | \vec{k} + \vec{K}' \rangle \tag{14.29}$$

$$S_{\vec{K}\vec{K}'} \equiv \langle \vec{k} + \vec{K} | \vec{k} + \vec{K}' \rangle \tag{14.30}$$

Exercise 14.7   Show that the matrix elements and overlap integrals do not depend explicitly on $\vec{k}$.

Using Eq. 14.27, we can show that

$$S_{\vec{K}\vec{K}'} = \langle \vec{K} | \vec{K}' \rangle = \delta_{\vec{K}\vec{K}'} \tag{14.31}$$

where $\delta_{\vec{K}\vec{K}'}$ is the Kronecker delta function. In addition,

$$\mathscr{H}_{\vec{K}\vec{K}'} = \langle \vec{k} + \vec{K} | - \frac{\hbar^2}{2m} \nabla_{\vec{k}}^2 | \vec{k} + \vec{K}' \rangle + \langle \vec{k} + \vec{K} | U(\vec{r}) | \vec{k} + \vec{K}' \rangle \tag{14.32}$$

where $U(\vec{r})$ is a weak, periodic potential. The kinetic energy eigenvalues are easy to find and, of course, are identical with the free-electron eigenvalues. The potential energy term is more problematic. However, since $U(\vec{r})$ is periodic, we can expand it in a Fourier series,

$$U(\vec{r}) = \sum_{\vec{K}''} U_{\vec{K}''} |\vec{K}''\rangle \tag{14.33}$$

The coefficients $U_{\vec{K}''}$ are called the structure factors of our crystal. They have the property that $U_{-\vec{K}} = U_{\vec{K}}^*$. Thus,

$$\mathscr{H}_{\vec{K}\vec{K}'} = \frac{\hbar^2}{2m} (\vec{k} + \vec{K}')^2 \delta_{\vec{K}\vec{K}'} + \frac{1}{V^{\frac{1}{2}}} \sum_{\vec{K}''} U_{\vec{K}''} \delta_{\vec{K},\vec{K}'+\vec{K}''} \tag{14.34}$$

In principle, Eq. 14.28 with Eqs. 14.31 and 14.34 constitute the complete statement of the problem of electrons in a weak, periodic potential. We could feed the secular determinant to a computer and solve for the eigenvalues and eigenvectors. Note, however, that there are infinitely many reciprocal lattice vectors, so that the secular determinant is actually of infinite dimensions. For each given $\vec{k}$ value, the solutions are the different bands and we label them with a band index, n, as we have done for the tight-binding case. We can restrict the wavevector $\vec{k}$ to lie in the first Brillouin zone.

In practice, of course, it is somewhat difficult to solve an infinitely large determinant. What can be done instead is to solve only for the lowest-energy roots desired. This is similar to limiting the tight-binding basis to, for example, only the occupied atomic orbitals. In actual fact, interesting things happen only for selected reciprocal lattice vectors. We will examine these more restricted cases now.

Exercise 14.8   Under what conditions does the nearly-free-electron problem reduce to the free-electron problem?

The first case that we examine is an extreme one, but it reveals some interesting properties and insights. Let us consider only two reciprocal lattice vectors in the expansion for the wavefunction (Eq. 14.27), one of which is $\vec{K}=0$, the other being still arbitrary and labeled $\vec{K}$. Then, the secular determinant contains only four terms,

$$\begin{aligned} S_{00} &= S_{\vec{K}\vec{K}} = 1 \\ S_{0\vec{K}} &= S_{\vec{K}0} = 0 \end{aligned} \tag{14.35}$$

$$\begin{aligned} \mathcal{H}_{00} &= \mathcal{E}_{\vec{k}} + \frac{1}{\sqrt{V}}U_0 \\[2mm] \mathcal{H}_{\vec{K}\vec{K}} &= \mathcal{E}_{\vec{k}+\vec{K}} + \frac{1}{\sqrt{V}}U_0 \\[2mm] \mathcal{H}_{0\vec{K}} &= \frac{1}{\sqrt{V}}U_{\vec{K}}^* \\[2mm] \mathcal{H}_{\vec{K}0} &= \frac{1}{\sqrt{V}}U_{\vec{K}} \end{aligned} \tag{14.36}$$

Exercise 14.9   Verify that Eqs. 14.35 are the correct overlap integrals and that Eqs. 14.36 are the correct matrix elements.

Therefore, the secular determinant is

$$\begin{vmatrix} \mathcal{E}_{\vec{k}} + \frac{1}{\sqrt{V}}U_0 - \mathcal{E} & \frac{1}{\sqrt{V}}U_{\vec{K}}^* \\[2mm] \frac{1}{\sqrt{V}}U_{\vec{K}} & \mathcal{E}_{\vec{k}+\vec{K}} + \frac{1}{\sqrt{V}}U_0 - \mathcal{E} \end{vmatrix} = 0$$

Defining $\tilde{\mathcal{E}} \equiv -(1/\sqrt{V})U_0 + \mathcal{E}$ and expanding the determinant gives

$$\mathcal{E}_{\vec{k}}\mathcal{E}_{\vec{k}+\vec{K}} - \tilde{\mathcal{E}}(\mathcal{E}_{\vec{k}} + \mathcal{E}_{\vec{k}+\vec{K}}) + \tilde{\mathcal{E}}^2 - \frac{1}{V}|U_{\vec{K}}|^2 = 0 \tag{14.37}$$

Exercise 14.10   Obtain Eq. 14.37.

This is a quadratic equation in $\tilde{\mathscr{E}}$ and it has two roots,

$$\tilde{\mathscr{E}}_{\pm} = \frac{1}{2}(\mathscr{E}_{\vec{k}} + \mathscr{E}_{\vec{k}+\vec{K}}) \pm \frac{1}{2}((\mathscr{E}_{\vec{k}} + \mathscr{E}_{\vec{k}+\vec{K}})^2 - 4\mathscr{E}_{\vec{k}}\mathscr{E}_{\vec{k}+\vec{K}} + \frac{4}{V}|U_{\vec{K}}|^2)^{\frac{1}{2}} \quad (14.38)$$

Exercise 14.11 Show that Eq. 14.38 is correct.

Two nondegenerate bands arise from allowing interaction between the plane waves. This is a rather complicated-looking expression, but let us examine it in detail. Firstly, the energy difference between the bands $\tilde{\mathscr{E}}_{+}$ and $\tilde{\mathscr{E}}_{-}$ is

$$\Delta\tilde{\mathscr{E}} = ((\mathscr{E}_{\vec{k}} - \mathscr{E}_{\vec{k}+\vec{K}})^2 + \frac{4}{V}|U_{\vec{K}}|^2)^{\frac{1}{2}} \quad (14.39)$$

Recall that the free-electron-state energies are $\mathscr{E}_{\vec{k}}$ and $\mathscr{E}_{\vec{k}+\vec{K}}$. The free-electron energies are replotted in Fig. 14.6, where now we have included the underlying periodic arrangement of atoms in the crystal. For the sake of simplicity, we draw the band structure in just one dimension. The first Brillouin zone boundaries are, as before, $k = \pm\pi/a$. The free-electron band structure plotted as a solid line continues outside the first Brillouin zone. This is known as the *extended zone scheme*. Recall, however, from Ch. 8 that only wavevectors within the first Brillouin zone are unique. Any wavevector outside the first Brillouin zone can be translated back by suitable addition or subtraction of a reciprocal lattice vector. By this process we recover the *reduced zone scheme*. The translated free-electron bands are shown as dashed lines in Fig. 14.6. The dashed band is also known as the second free-electron band.

Now, if in Eq. 14.39, $(\mathscr{E}_{\vec{k}} - \mathscr{E}_{\vec{k}+\vec{K}}) \gg (2/\sqrt{V})|U_{\vec{K}}|$, then the difference in energy between the two nearly-free-electron bands is given by $(\mathscr{E}_{\vec{k}} - \mathscr{E}_{\vec{k}+\vec{K}})$. But this is just the free-electron band energy difference! Mixing two plane waves has had little effect on the band structure. When will the above inequality be true? From Fig. 14.7, in which we replot the reduced zone free-electron band structure of Fig. 14.6, $(\mathscr{E}_{\vec{k}} - \mathscr{E}_{\vec{k}+\vec{K}})$ is large whenever $\vec{k}$ is *not* near a zone boundary. One choice is depicted in Fig. 14.7.

However, if we choose a wavevector closer to the zone boundary, the energy difference between free-electron bands diminishes and the correction term in Eq. 14.39 becomes significant. If we choose $\vec{k}$ right at the zone boundary, then $\mathscr{E}_{\vec{k}} = \mathscr{E}_{\vec{k}+\vec{K}}$ and $\Delta\mathscr{E} = (2/\sqrt{V})|U_{\vec{K}}|$. What is the meaning of this energy difference? Simply put, because of the mixing of different plane waves, which is a consequence of the weak, periodic potential, the nearly-free-electron bands *split* at the zone boundary. In other words, a band gap opens up! The nearly-free-electron bands are sketched on Fig. 14.7.

What is the physical significance of this band gap and what determines

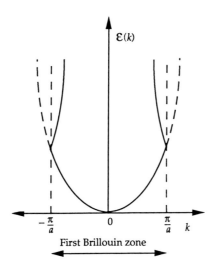

Fig. 14.6.   The free-electron band structure for a one-dimensional system in the extended zone scheme. To obtain the reduced zone scheme, bands outside the first Brillouin zone are translated back into the first zone by shifting them by a reciprocal lattice vector. These bands are shown by dashed lines.

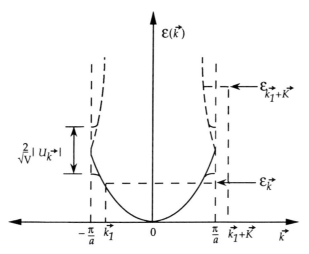

Fig. 14.7.   Lowest-energy free-electron bands (solid lines) and free-electron bands from the second Brillouin zone (dashed lines). A choice of $\mathscr{E}_{\vec{k}}$ for which plane-wave mixing has little effect on the band structure is shown. Splitting of the bands occurs for wavevectors near the zone boundary.

its magnitude? If $\vec{k}$ is very close to a zone boundary, then it lies in a *Bragg plane*, *i.e.*, the plane defined by $\pm\frac{1}{2}|\vec{K}|$ in the one-dimensional case. Bragg planes are planes for which incident beams of X-rays can be reflected or diffracted.

Recall that we can interpret $\hbar k$ as the momentum of a free electron. De Broglie related the electron momentum to a wavelength,

$$p = \frac{h}{\lambda} \tag{14.40}$$

so that $\lambda = 2\pi/k$. For a wavevector at the zone boundary, $k = \pi/a$ and $\lambda = 2a$. A typical interatomic spacing is $a \approx 2$ Å , from which $\lambda \approx 4$ Å, a wavelength in the X-ray part of the electromagnetic spectrum.

The wavefunctions for wavevectors near the zone boundary are easily obtained for the two bands $\tilde{\mathscr{E}}_{\pm} = \frac{1}{2}(\mathscr{E}_{\vec{k}} + \mathscr{E}_{\vec{k}+\vec{K}}) \pm (1/\sqrt{V})|U_{\vec{K}}|$. From the secular determinant, we find that the coefficients satisfy

$$c_0 = \pm(\text{sign } U_{\vec{K}})c_{\vec{K}} \tag{14.41}$$

where the sign of $U_{\vec{K}}$ is undetermined. We obtain two sets of solutions. If $U_{\vec{K}} > 0$,

$$\tilde{\mathscr{E}}_{+} = \frac{1}{2}(\mathscr{E}_{\vec{k}} + \mathscr{E}_{\vec{k}+\vec{K}}) + \frac{1}{\sqrt{V}}|U_{\vec{K}}| \qquad |\psi(\vec{r})|^2 \propto (\cos(\tfrac{1}{2}\vec{K}\cdot\vec{r}))^2$$

$$\tilde{\mathscr{E}}_{-} = \frac{1}{2}(\mathscr{E}_{\vec{k}} + \mathscr{E}_{\vec{k}+\vec{K}}) - \frac{1}{\sqrt{V}}|U_{\vec{K}}| \qquad |\psi(\vec{r})|^2 \propto (\sin(\tfrac{1}{2}\vec{K}\cdot\vec{r}))^2 \tag{14.42}$$

and if $U_{\vec{K}} < 0$,

$$\tilde{\mathscr{E}}_{+} = \frac{1}{2}(\mathscr{E}_{\vec{k}} + \mathscr{E}_{\vec{k}+\vec{K}}) + \frac{1}{\sqrt{V}}|U_{\vec{K}}| \qquad |\psi(\vec{r})|^2 \propto (\sin(\tfrac{1}{2}\vec{K}\cdot\vec{r}))^2$$

$$\tilde{\mathscr{E}}_{-} = \frac{1}{2}(\mathscr{E}_{\vec{k}} + \mathscr{E}_{\vec{k}+\vec{K}}) - \frac{1}{\sqrt{V}}|U_{\vec{K}}| \qquad |\psi(\vec{r})|^2 \propto (\cos(\tfrac{1}{2}\vec{K}\cdot\vec{r}))^2 \tag{14.43}$$

These two sets of wavefunctions are plotted in Fig. 14.8. The two solutions are sometimes called *s-like* ($\cos^2(\tfrac{1}{2}\vec{K}\cdot\vec{r})$) and *p-like* ($\sin^2(\tfrac{1}{2}\vec{K}\cdot\vec{r})$) because of the positions of the nodes in the charge densities.

Exercise 14.12 Find Eqs. 14.42 and 14.43.

Earlier, in Ch. 12, we discussed how band gaps come about because of local chemical interactions. In particular, it was the difference in energy

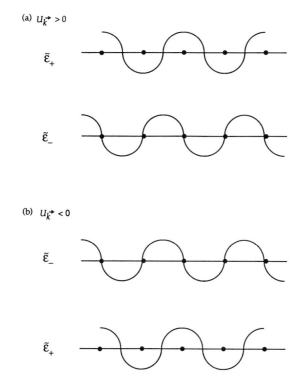

Fig. 14.8. The wavefunctions for the case of an electron in a weak periodic potential. (a) $U_{\vec{K}} > 0$ and (b) $U_{\vec{K}} < 0$.

of the atomic orbitals that caused gaps in the band structure. The nearly-free-electron approach seems to be saying something quite different: namely, that band gaps arise from the interaction of free electrons with the periodic potential of a crystal. The two viewpoints are not orthogonal, however. It is the potential energy of the individual *atoms* that provides the periodic crystal potential. In an isolated atom, this same potential energy gives rise to atomic orbitals with nondegenerate energies. Energy bands are determined by the competition *between* the potentials of individual atoms and the overlap of their orbitals. The free-electron model works best when the overlap is very strong, so that the atomic orbitals tend to lose their identity. This is particularly true for elements in columns I, II, and III of the periodic table. We can treat covalent and ionic systems with the machinery of the free- and nearly-free-electron approaches. However, a much stronger periodic potential is required. It is easiest, therefore, to think of these solids in terms of the overlap of atomic orbitals. Finally, if we compare the band structure for the free-electron system (or even the nearly-free-electron system for small $|\vec{k}|$) with the band structure of crystalline Si (Ch. 13), we see some striking similarities. The bottom of the Si bands at the Γ point looks remarkably similar to the bottom of the free-electron band. We also showed earlier how the band structure $\mathscr{E}(\vec{k})$ can be expanded for small k to yield a quadratic dependence of $\mathscr{E}$ on k.

Exercise 14.13 As a further manifestation of the similarity between free-electron bands, nearly-free-electron bands, and tight-binding bands, calculate the effective mass of an electron in each of these types of bands. For what values of wavevector do the masses coincide? Where are the greatest discrepancies?

Exercise 14.14 Show that the free-electron wavefunction can be written in the form of a Bloch wavefunction. Hint: Use the fact that the wavevector is periodic in the reciprocal lattice vector. What is the form of the periodic function in the Bloch wavefunction so found?

Problems

14.1 Exercise 14.2.

14.2 Exercise 14.3.

14.3 Exercise 14.6.

14.4 Exercise 14.13.

14.5 Exercise 14.14.

14.6 The *structure factor* of a crystal structure is given by

$$U_{\vec{K}} = \sum_j f_j \exp(-\imath \vec{K} \cdot \vec{r}_j) \tag{P14.1}$$

where the sum is over all atoms j in a single unit cell, $\vec{K}$ is a reciprocal lattice vector, f_j is the atomic form factor, and $\vec{r}_j$ is a vector to each atom in the unit cell. Consider the $AuCu_3$ system, which may be ordered or disordered.

(a) For the *completely disordered* system, calculate the structure factor. Hint: The unit cell can be thought of as an fcc cell in which each site is occupied by an 'average' gold–copper atom. There are four atoms per unit cell.

(b) For the *completely ordered* system, there is one Au atom at $(0,0,0)$ and one Cu atom each at $(\frac{1}{2}, \frac{1}{2}, 0)$, $(\frac{1}{2}, 0, \frac{1}{2})$, and $(0, \frac{1}{2}, \frac{1}{2})$. See Fig. 14.9. Calculate the structure factor for the completely ordered crystal.

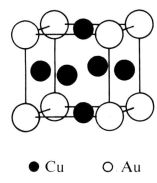

● Cu   ○ Au

Fig. 14.9. The $AuCu_3$ crystal structure.

    (c) Can you distinguish between the completely ordered and completely disordered crystals? How?

14.7 Consider an AB *random crystalline* alloy in which the coordination number is z. By random we mean that the A and B atoms are distributed in crystalline sites randomly. The valence states of each type of atom are assumed to be s states. Assume that the nearest-neighbor hopping integrals are always β, independent of the actual atoms that the electron is hopping between. Let the on-site matrix elements be $\mathscr{E}_A$ and $\mathscr{E}_B$ for atoms A and B, respectively, and let $\mathscr{E}_B > \mathscr{E}_A$. Show that an energy gap exists if

$$\mathscr{E}_B - \mathscr{E}_A > 2z|\beta| \qquad\qquad (\text{P14.2})$$

Hint: Remember that the wavefunctions of this system will just be linear combinations of the atomic orbitals on each atom. Put the expansion for the wavefunction into the Schrödinger equation to find an expression for $|\mathscr{E} - \mathscr{E}_j|$, where j is an arbitrary atomic site.

14.8 Construct the first six free-electron bands for an fcc lattice along the [100] direction in the Brillouin zone from the zone center to the zone boundary.

14.9 For a free electron gas at $T = 0$ K with a fixed number of electrons occupying a volume V and with a Fermi energy $\mathscr{E}_F$, find:

    (a) the average electronic kinetic energy;
    (b) an expression for $\partial\mathscr{E}_F/\partial V$;
    (c) the bulk modulus;
    (d) the pressure of the electron gas;
    (e) how the average electronic kinetic energy changes if $T > 0$ K.

14.10 We have calculated the band structure for a one-dimensional nearly-free-electron metal in the text. Consider now the same problem in a two-dimensional system with square-planar symmetry.

(a) Calculate the band energies for a model system in which only two reciprocal lattice vectors are considered. Plot the band structure around the circuit $\Gamma \to X \to M \to \Gamma$ in the first Brillouin zone for *three* different cases: (1) zero periodic potential (= constant); (2) weak periodic potential; and (3) strong periodic potential. Be certain to take into account any symmetry constraints on the bands.

(b) For each of the three cases considered in (a) above, draw a schematic of the accompanying DOS.

(c) For each of the three cases considered in (a) above, sketch the curves of constant energy in the first Brillouin zone.

Note that when the Fermi surface touches a Brillouin zone boundary, $\mathcal{E}_{\vec{k}}$ flattens out and this gives a peak in the DOS followed by a drop. Different crystal structures in three dimensions give rise to differently shaped Brillouin zones and thus the number of electrons per atom, N, required to touch the zone face will vary.

A rough but simple description of the alloying process is provided by the *rigid-band model*. Consider the problem of alloying Zn with Cu to form brass alloys. We begin with pure Cu, which has the fcc crystal structure with one valence $4s$ electron per atom (and one atom per unit cell). Zn atoms are randomly added, but we assume that the band structure remains identical with that of pure Cu. Each Zn atom adds two valence electrons. Jones (1934) predicted a change in the stability of various crystal structures as more and more Zn is added and the electron/atom ratio increases. These changes in structure were first noted by Hume-Rothery in 1926.

(d) Calculate the electron per atom ratio for the fcc, bcc, and hexagonal structures at which the Fermi surface first touches the zone boundary. You may assume a free-electron model.

(e) Using your results of (d) above, as well as any insights from the first three parts of this problem, predict the structures of the alloys CuZn, Cu_5Zn_8, and $CuZn_3$.

Jones' simple interpretation of the Hume-Rothery rules is very appealing, but unfortunately wrong, because it oversimplifies the DOS (which is a result of the simplifications made in the free-electron and nearly-free-electron pictures).

14.11 Consider the following procedure for constructing a grain boundary. Take a cube of *cubic* material in which the crystallographic axes $< 100 >$ coincide with the sample's edges (*i.e.*, the [100] axis is the same as one of the cube's edges). Slice this cube in half along a (100) plane and rotate *one* of the halves by 30° either clockwise or counterclockwise about an axis that is *perpendicular* to the (100) plane exposed. Rebond the two halves to create a *bicrystal* which now has a *twist* grain boundary.

Assume that you have made twist grain boundaries with two different materials, first with Al and then with NaCl. Do not worry about the details of whether such a procedure is experimentally feasible or not.

(a) Consider bonding at the Al grain boundary first. You will have two different on-site matrix elements corresponding to the *s* and *p* valence electrons in addition to the three hopping integrals corresponding to the *s–s*, *s–p*, and *p–p* hopping between nearest neighbors. Determine the sign of these three hopping integrals. Using the moments theorem, discuss the nature of bonding in this grain boundary, *i.e.,* what kind of structure do you expect?

(b) Next consider bonding at the NaCl grain boundary. We have two distinct on-site matrix elements, corresponding to the Na site and the Cl site, and three possible hopping integrals, β_{Na-Na}, β_{Na-Cl}, and β_{Cl-Cl}. What are the signs and the relative magnitudes of these three hopping integrals? Using these and the moments theorem, discuss the nature of bonding in this grain boundary.

(c) Compare and contrast the two boundaries' structures and electronic structures.

15

Transition metals

15.1 Introduction

So far we have examined the electronic structure of systems that had s or p (or both) valence electrons. When we turn to the transition series, we must contend with valence d orbitals as well. The transition metals present perhaps the greatest challenge to theory, for these materials can show magnetic moments, a wide variety of structures, and unusual trends in the Fermi energy and cohesive energy across the transition series, and simple band-filling arguments for the conductivity can break down.

The most important aspect contributing to the transition metals' unusual properties is the relative extent of the valence d orbital wavefunctions versus that of the valence s and p orbital wavefunctions. From Fig. 15.1, we can see that the $3d$ atomic orbital is contracted relative to the $4s$ atomic orbital. In turn, then, the overlap between $3d$ orbitals on neighboring atoms is small compared with the overlap between neighboring $4s$ and $4p$ orbitals. But, as we first saw in Ch. 8, small overlap leads to narrow bands. The DOS of transition metals is thus characterized by narrow d bands superimposed on much broader s–p bands. A schematic DOS is shown in Fig. 15.2. The d band has a high density of states because a total of ten electrons can be accommodated within a narrow energy range. We expect the Fermi energy to be near the bottom of the d band for early transition metals and to move steadily upward through the d band across the series. Since the states at the bottom of the band are bonding in character, while those at the top are antibonding, we expect the bond energy to be a maximum in the middle of the transition series. A plot of the sublimation energy, which is the energy required to go from the solid state to the gaseous state, is shown in Fig. 15.3. The $4d$ and $5d$ transition elements do indeed show a parabolic variation of the sublimation energy. Surprisingly, the $3d$ series shows a *dip* in the

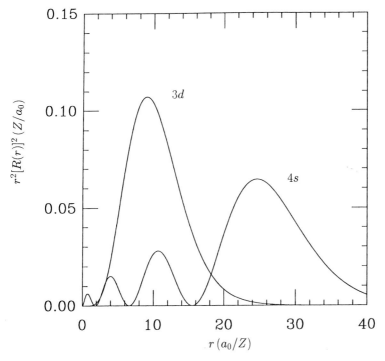

Fig. 15.1. Plot of the radial dependence of the 4*s* atomic wavefunction compared with the 3*d* atomic wavefunction.

middle of the series. Furthermore, the magnitude of the sublimation energy is smaller for the 3*d* series than for either the 4*d* or 5*d* series. The difference in magnitude is a consequence of the contracted radial extent of the 3*d* atomic orbitals. The 4*d* and 5*d* orbitals are not so severely contracted, and the *d* bands in the DOS for these two series are broader.

But why is there a dip in the energies of sublimation at the center of the 3*d* transition series? This feature has an interesting connection with the magnetic properties of these metals. Following Hund's rules for atoms, we occupy the 3*d*-derived states so as to maximize spin. In solids the same rules apply, but other effects are important. In particular, we know that bond energy is gained by *pairing electrons* in the same state. In the 3*d* transition series, however, bonding is weak. Therefore, it is sometimes more energetically favorable to put electrons in higher-energy, antibonding states with parallel spins in the *d* band than to incur the energy penalty of antiparallel spin electrons. The sublimation energies fall at the center of the 3*d* transition series because antibonding orbitals are occupied. Such effects

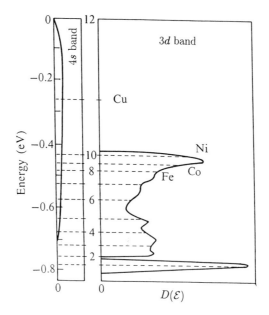

Fig. 15.2. Schematic density of states for a transition metal crystal. The 3*d* and 4*s* bands are shown separated. The dashed lines indicate the highest filled levels for Fe, Co, Ni, and Cu, assuming that the other transition metals have the same density of states as Cu. After H.M. Krutter, *Phys. Rev.* **48**, 664 (1935), J. C. Slater, *J. Appl. Phys.* **8**, 385 (1937), and G.F. Koster, *Phys. Rev.* **98**, 901 (1955).

are not seen in the 4*d* and 5*d* transition metals because the overlap is larger and, hence, more net bonding. Furthermore, the abundance of unpaired spins in the 3*d* transition series gives rise to a net magnetic moment for these materials, an effect which is absent in the 4*d* and 5*d* transition metals.

Crystal structures for the 4*d* and 5*d* transition series show a systematic variation as the *d* band is filled. The changes in structure are shown in Fig. 15.4. Again, the magnetism present in the 3*d* series complicates any clear variation in crystal structures.

We discussed the simple relationship between band filling and conductivity in Ch. 12. There we pointed out that any system with partially filled bands should be metallic. There are five *d* atomic orbitals, but inspection of the character tables (see Appendix 3) shows that there are no point groups† which have five-fold degenerate irreducible representations. Therefore, the *d* orbitals *must split* into levels that are at most three-fold degenerate. In this case, the lower-energy *d* bands may be filled with a gap before the higher-energy unfilled bands.

† The icosahedral group is an exception.

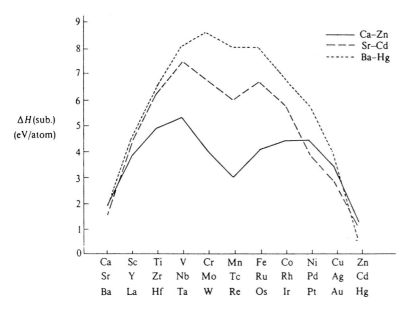

Fig. 15.3. Sublimation energies for transition metals of the three series. Used with permission from P.A. Cox, *The Electronic Structure and Chemistry of Solids* (Oxford University Press, Oxford, 1987), p. 69.

4*d* series

Y	Zr	Nb	Mo	Tc	Ru	Rh	Pd	Ag	Cd
hcp	hcp	bcc	bcc	hcp	hcp	fcc	fcc	fcc	hcp

5*d* series

La	Hf	Ta	W	Re	Os	Ir	Pt	Au	Hg
hex	hcp	bcc	bcc	hcp	hcp	fcc	fcc	fcc	rhomb

Fig. 15.4. Trends in the crystal structures for elemental transition metals of the 4*d* and 5*d* series.

A number of nonmetallic transition metal compounds cannot be understood with such a simple model, however. NiO is perhaps the most famous case (or infamous, depending on one's viewpoint). The electronic structure of atomic Ni is $[Ar]3d^84s^2$. In NiO, each Ni^{2+} ion has two unpaired electrons, making the material magnetic. NiO cannot have a completed filled d band and our simple arguments from Ch. 12 indicate that it should be metallic. Most surprisingly, however, NiO is a good insulator!

The complete failure of band theory to predict the correct electrical properties of NiO caused quite a bit of consternation in the solid state physics community in the years immediately preceding the Second World War and for many years afterwards. Entire conferences and books were dedicated to unraveling the 'NiO' problem. The reason for the failure of band theory can again be traced to the relative narrowness of the d bands *and* our neglect of electron repulsion. Electron repulsion keeps electrons away from one another, localized on individual atoms. The weak overlap between $3d$ atomic orbitals provides little incentive for band formation in this series (and hence the narrow d bands). In the $4d$ and $5d$ transition metals, however, the d orbitals are more extended. They can more readily form bonds and overcome the effects of electron repulsion.

In this chapter we introduce a model for describing properties of the transition metals based on a tight-binding picture due to Friedel. We also examine the types of bonds possible when d atomic orbitals overlap and the relative stability of the various crystal structures for elemental transition metals. We conclude the chapter with a discussion of some case studies: first, elemental copper, followed by a transition metal oxide, RuO_2.

15.2 The Friedel model

Friedel[†] assumed that the variation in properties across the transition metal series arose from filling the d band. He represented the d band density of states as a rectangle of width W and center of gravity $\mathcal{E}_d$. Because the completely filled band contains ten electrons, the rectangle's height is $5/W$. See Fig. 15.5.

The trends observed for the sublimation energies for the $4d$ and $5d$ transition series can be explained by examining the bond energy. Recall from Ch. 7 that the bond energy is the 'glue' that holds atoms together in a solid. The *bond* energy per atom of the d band within the Friedel model is given

---

[†] J. Friedel in *Physics of Metals*, ed. J. Ziman (Cambridge University Press, London, 1969), p. 340.

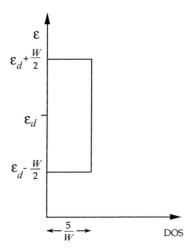

Fig. 15.5. Density of states assumed in the Friedel model.

by

$$\mathscr{E}_{\text{bond}} = 2\int_{\mathscr{E}_d - \frac{W}{2}}^{\mathscr{E}_F} (\mathscr{E} - \mathscr{E}_d)\frac{5}{W}\,\mathrm{d}\mathscr{E} \tag{15.1}$$

where the factor of 2 accounts for the two possible spin orientations. Performing the straightforward integration, we find

$$\mathscr{E}_{\text{bond}} = -\frac{W}{20}N_d(10 - N_d) \tag{15.2}$$

where N_d is the number of d electrons per atom,

$$N_d = 2\int_{\mathscr{E}_d - \frac{W}{2}}^{\mathscr{E}_F} \frac{5}{W}\,\mathrm{d}\mathscr{E} = \frac{10}{W}(\mathscr{E}_F - \mathscr{E}_d + \frac{W}{2}) \tag{15.3}$$

$\mathscr{E}_d$ is the center of gravity and W is the band width. The variation of $\mathscr{E}_{\text{bond}}$ with N_d is plotted in Fig. 15.6. The bond energy is maximized for a half-filled band and varies parabolically with d band filling; this is in good accord with the trends in sublimation energy for the $4d$ and $5d$ transition series. However, as we noted above, this simple model neglects electron repulsion effects and cannot correctly predict properties of the $3d$ metals. We turn to this problem next.

15.3 The moments theorem and the Friedel model

The equilibrium structure and cohesive energy for transition metals is, in the simplest approximation, correctly given by including both bond energy,

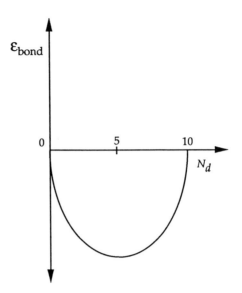

Fig. 15.6. Variation of the bond energy with number of *d* electrons.

$\mathscr{E}_{bond}$, and electron repulsion, $\mathscr{E}_{rep}$, terms. The cohesive energy for the solid, $\mathscr{E}_{coh}$, which is a measure of how much energy is gained when the atoms condense from infinitely far away, can thus be written as

$$\mathscr{E}_{coh} = \mathscr{E}_{bond} + \mathscr{E}_{rep} \tag{15.4}$$

The bond energy we have calculated above. The electron repulsion term is assumed to arise from a two-atom interaction and is represented by what is called a 'pair potential'. We want to examine the relationship between the width of the density of states within the Friedel model, W, and the local atomic environment. To do this we use the second moment of the DOS distribution function, $\mu^{(2)}$.

Now, we know from Ch. 9 that the second moment of the DOS is given by

$$\mu^{(2)} = \sum_{i \neq j} \mathscr{H}_{ji}\mathscr{H}_{ij} = z\beta^2 \tag{15.5}$$

where z is the coordination number and β is the first-neighbor hopping integral. But, from the rectangular DOS assumed in the Friedel model, we can write

$$\mu^{(2)} = \int_{\mathscr{E}_d - \frac{W}{2}}^{\mathscr{E}_d + \frac{W}{2}} (\mathscr{E} - \mathscr{E}_d)^2 \frac{1}{W} \, d\mathscr{E} = \frac{W^2}{12} \tag{15.6}$$

Equating Eqs. 15.5 and 15.6,

$$W = \sqrt{12z}|\beta| \qquad (15.7)$$

so that, as we discussed previously, the band width is proportional to the square root of the coordination and the magnitude of the hopping integral. We can continue further with this analysis if we inquire about the functional dependence of the hopping integral on the interatomic bond length, R. We found in Ch. 6 that the radial solutions to the Schrödinger equation for the H atom decay exponentially away from the atom. Therefore, let us assume the form

$$\beta(R) = bN_d e^{-\lambda R} \qquad (15.8)$$

where b and λ are unknown constants and we have also assumed that the hopping integral is proportional to the number of d electrons per atom.

The repulsive interaction, $\phi(R)$, between neighboring atoms originates from the Coulomb repulsion between neighboring charge densities, and we anticipate that this interaction will vary as the square of β,

$$\begin{aligned} \phi(R) &= [\beta(R)]^2 \\ &= aN_d^2 e^{-2\lambda R} \end{aligned} \qquad (15.9)$$

where a is an unknown constant.

Substituting Eqs. 15.2 and 15.9 into Eq. 15.4, the cohesive energy per atom is

$$\mathscr{E}_{coh} = \frac{z}{2}aN_d^2 e^{-2\lambda R} - \frac{W}{20}N_d(10 - N_d) \qquad (15.10)$$

and substituting Eq. 15.7 for W gives

$$\mathscr{E}_{coh} = \frac{z}{2}aN_d^2 e^{-2\lambda R} - \frac{bN_d\sqrt{12z}}{20}e^{-\lambda R}N_d(10 - N_d) \qquad (15.11)$$

Equation 15.11 is plotted in Fig. 15.7 as a function of the interatomic distance, R. The minimum energy of this curve determines the equilibrium bond distance in a transition metal crystal. Assuming only nearest-neighbor interactions, we can show that the equilibrium bond distance is

$$R_0 = \frac{1}{\lambda}\ln\left[\frac{10a\sqrt{z}}{\sqrt{3}b(10 - N_d)}\right] \qquad (15.12)$$

the equilibrium cohesive energy is

$$\mathscr{E}_{coh} = -\frac{3b^2}{200a}[N_d(10 - N_d)]^2 \qquad (15.13)$$

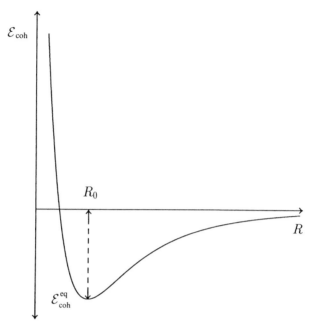

Fig. 15.7.   Variation of the cohesive energy with interatomic separation. The equilibrium cohesive energy, $\mathscr{E}_{coh}^{eq}$, occurs at the equilibrium interatomic separation, R_0.

the band width is

$$W = \frac{3b^2}{5a}N_d(10 - N_d) \tag{15.14}$$

and the bulk modulus is

$$B = \frac{2\lambda^2}{9R_o}\mathscr{E}_{coh} \tag{15.15}$$

Numerical values can be found for all these properties by fitting the equilibrium interatomic distance to experimental values and by fitting the ratio b^2/a to the experimental band width.

Exercise 15.1   Derive Eqs. 15.12 through 15.15.

The predicted cohesive energy is about 2 eV smaller than the experimental value. This discrepancy arises primarily from neglect of the bonding from *s* and *p* valence electrons. Inclusion of the extra valence electrons increases the cohesive energy, so at least the trend is correct.

One of the intriguing results from the simple Friedel model in the second moment approximation is that the cohesive energy does not depend on the atomic coordination, z. Physically this means that the cohesive energy is independent of the crystal structure. To find equilibrium crystal structures, we must include higher moments of the DOS. Lastly, note that the bond energy varies as the square root of the coordination rather than linearly, which is a reflection of the unsaturated nature of the metallic bond. By 'unsaturated' we mean that there are fewer valence electrons than there are nearest neighbors, so that it is difficult to isolate individual bonds in a metallic solid.

15.4 Nature of bonding in systems with d electrons

When we discussed bonding in diatomic molecules in Ch. 7, we pointed out the types of bonding arrangements that arise from overlap of s and p atomic orbitals. There we saw that two kinds of bonds could be formed, σ bonds and π bonds. Recall that with the five available d orbitals, three types of bond can be formed. These are called σ, π, and δ. The σ bond arises from an interaction between d_{z^2} atomic orbitals which lie along the same axis. The bond is denoted $(dd\sigma)$ and is shown in Fig. 15.8. π-type bonds, in which the charge density has a node in the plane directly between the neighboring atoms, can be formed from the overlap of any of the d_{xy}, d_{yz}, or d_{xz} orbitals, provided they are appropriately oriented. $(dd\pi)$ bonding combinations are shown in Fig. 15.9. As for σ and π bonds derived from s and p valence electrons, we expect the σ bond energy to be larger than the π bond energy for the d valence electrons. In addition, we expect the $(dd\sigma)$ and $(dd\pi)$ bonds to be weaker than their s and p counterparts. The weakest category of bond formed in this scheme is the $(dd\delta)$ bond and it arises from the overlap of neighboring $d_{x^2-y^2}$ orbitals and, independently, from one of the orbitals of the type d_{xy}, depending on the orientation of the coordinate system. An example of a δ bond is shown in Fig. 15.10.

The $(dd\sigma)$ hopping integral is negative because the lobes that overlap have the same sign.† Similarly, the $(dd\delta)$ hopping integral is also negative. However, the $(dd\pi)$ hopping integral is positive.

A case in which δ bonds provide important structure stabilization is the Chevrel phase compound $PbMo_6S_8$. The structure consists of a cubic ordered arrangement of Mo_6S_8 clusters. The clusters themselves are of cubic symmetry, but they do not remain parallel to one another in the equilibrium structure of the solid. Owing to the formation of δ bonds by overlap between empty d orbitals on Mo and filled orbitals on S, the structure is stabilized, with the cubic clusters tilted slightly. This structure is shown in Fig. 15.11.

† Refer to the arguments given in Ch. 7.

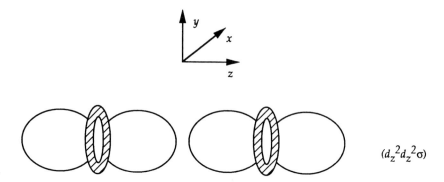

$(d_z{}^2 d_z{}^2 \sigma)$

Fig. 15.8. σ bond formed by overlap of two d_{z^2} atomic orbitals.

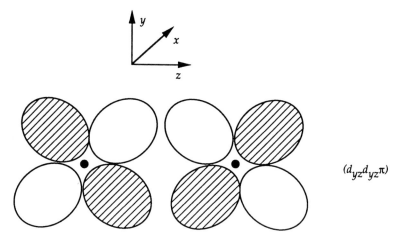

$(d_{yz} d_{yz} \pi)$

Fig. 15.9. π bond formed by overlap of two d_{yz} atomic orbitals that are appropriately oriented.

15.5 Equilibrium crystal structures

From the discussion of bonding within the Friedel model, we found that the second moment approximation cannot discriminate between different crystal structures because the cohesive energy does not depend on the coordination number. We mentioned in the Introduction part of this chapter that the structures of the $4d$ and $5d$ transition metals showed systematic variation as the d band filled. The $3d$ transition metals are complicated by the presence of magnetism and we will only discuss the structural stability trends for the $4d$ and $5d$ series.

Figure 15.12 shows the results of a tight-binding calculation of the d bond

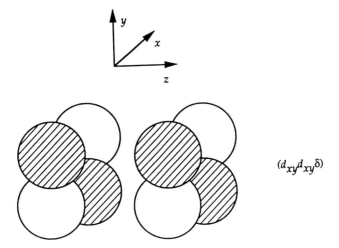

Fig. 15.10. δ bond formed by overlap of two d_{xy} atomic orbitals.

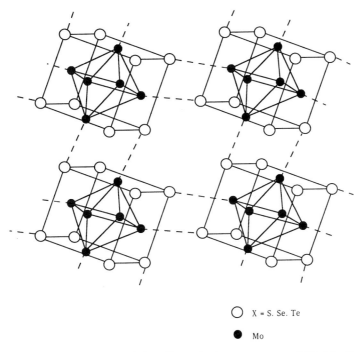

○ X = S. Se. Te

● Mo

Fig. 15.11. Crystal structure of the Chevrel phase compound $PbMo_6S_8$ in which δ bonding provides stabilization. After R. Hoffmann, *Solids and Surfaces: A Chemist's View of Bonding in Extended Structures* (VCH Publishers, Inc., New York, 1988), p. 80.

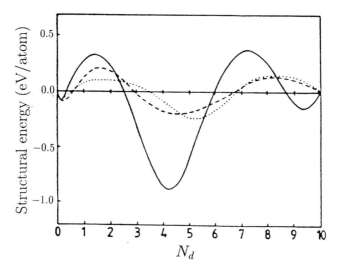

Fig. 15.12. Structural energies for a bcc (solid line), fcc (dashed line), and hcp (dotted line) crystal for a tight-binding d band calculation. Used with permission from D.G. Pettifor, *Solid State Physics*, **40**, 43 (1987).

energy for the bcc, fcc, and hcp crystal structures as a function of band filling. The curves predict that the structures will change in the order hcp → bcc → hcp → fcc → bcc as N_d varies from 0 to 10. This is in very good agreement with the actual structures observed. See Fig. 15.4. We therefore conclude that the changes seen in structural variation across the $4d$ and $5d$ transition metal series arise from filling of the d band. In order to understand this trend fully within the confines of our simple moments model, we need to calculate higher moments in the DOS. That is the subject of a more in-depth inquiry than we can undertake at present.

15.6 Case studies

15.6.1 Elemental copper

Elemental copper has an fcc lattice with a one-atom basis. The electron configuration of copper is $[Ar]4s^1 3d^{10}$. The Brillouin zone for copper, as constructed by the Wigner–Seitz method, is identical with that of silicon. This should reinforce for you the fact that the lattice determines the Brillouin zone, not the basis.

The band structure of copper is shown in Fig. 15.13. There are a number of interesting features in the figure. The lowest-energy valence band is at the Γ

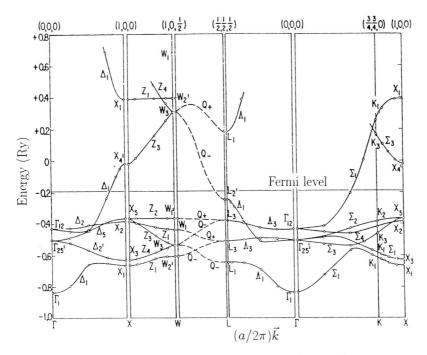

Fig. 15.13. The band structure of elemental copper along various symmetry lines in the Brillouin zone. See the text for a discussion. From B. Segall, *Phys. Rev.* **125**, 109 (1962). Used with permission.

point and, referring to Fig.14.2, looks remarkably free-electron-like. Indeed, its shape is nearly parabolic. Higher in energy, we come to a complex of narrow bands. This is the set of d-derived bands, although in copper there is significant mixing of s bands as well. The narrowness of the bands indicates the overlap is weak, which is a result of the contracted d orbitals. In turn, the effective mass, which varies inversely with band curvature, is very heavy. By examining a general k point along the Σ direction, we can see that the complex of narrow bands consists of five bands. The triply degenerate band at $\Gamma_{25'}$ splits, as does the doubly degenerate Γ_{12} band. The higher-energy valence band behaves like a free-electron band as we move away from the Γ point (band Σ_1).

The 11 valence electrons of copper can be accommodated in $5\frac{1}{2}$ bands. The Fermi level thus lies halfway up the top band, which is above the d bands. The electronic structure at $\mathscr{E}_F$ is free-electron-like.

Moving one element to the right, to zinc, the d bands narrow significantly

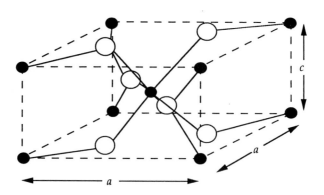

Fig. 15.14. The rutile structure adopted by RuO_2.

and drop in energy. Zinc behaves, to first approximation, as a free-electron metal with core d electrons.

15.6.2 RuO₂

Recall back in Ch. 2 we discussed MO_2 in the rutile structure, where M is a transition metal element. A schematic of the rutile structure is shown in Fig. 15.14. The Bravais lattice is primitive tetragonal and there are two formula units per unit cell. Atoms are located at:

$$Ru \qquad (0,0,0); (\tfrac{1}{2}, \tfrac{1}{2}, \tfrac{1}{2})$$
$$O \quad \pm(u,u,0); (u+\tfrac{1}{2}, \tfrac{1}{2}-u, \tfrac{1}{2})$$

$$(15.16)$$

Each ruthenium atom has six nearest neighbors, while each oxygen atom has coordination three. Many oxides of transition metals adopt the rutile structure (*e.g.*, Cr, Mn, Nd, Mo, Ta, W), for which $c \sim 4.6$ Å, $a \sim 3.0$ Å, and $u \sim 0.305$.

The electron configuration of ruthenium is $[Kr]4d^7 5s^1$ and that of oxygen is $[He]2s^2 2p^4$. We can therefore expect mixing between the metal s and d orbitals and the oxygen p orbitals. The oxygen s orbitals are generally too low in energy for appreciable mixing to occur. Since each ruthenium atom is in an octahedral environment, we know from the O_h character table (Appendix 3) that the d orbitals must split into two-fold and three-fold degenerate levels.

Exercise 15.2   From a sketch of the bonds, can you decide whether the three-fold degenerate level is higher or lower in energy than the two-fold degenerate level?

The Brillouin zone has tetragonal symmetry and it is shown in Fig. 15.15 with some of the special points labeled. Depending on how many of the atomic orbital interactions we include, we can expect anywhere from 24 bands (10 d functions, 2 s functions (Ru), and 12 p functions) to 26 bands (10 d functions, 12 p functions, and 4 s functions (O)) to 28 bands (10 d functions, 2 s functions (Ru), 12 p functions, and 4 s functions (O)).

The calculated band structure around a circuit of special points is shown in Fig. 15.16. The zero of energy is coincident with the top of the oxygen $2p$ bands. The d-derived bands lie above the complex of oxygen $2p$ bands. Notice how the d bands are broken up into a two-fold (e_g) and a three-fold (t_{2g}) set of bands which we anticipated on the basis of symmetry arguments about the Ru atomic environment. At very low energies lie the four oxygen s-derived bands; they are too low in energy to be included on this figure. Note that the lower-energy bands come from the more electronegative element, as we have anticipated before.

The Fermi energy, which is indicated by a dashed line, falls right in the middle of the d bands. Since $\mathscr{E}_F$ lies in the middle of a partially filled band, RuO_2 is metallic.

Exercise 15.3   How many electrons per unit cell are there in RuO_2?

Exercise 15.4   Will TiO_2 be an insulator, semiconductor, or conductor?

Problems

15.1   Consider a hypothetical solid with a simple cubic crystal structure and a one-atom basis. The valence electron configuration is [noble gas]$3s^n 3d^m$, where $0 \le n \le 10, 0 \le m \le 2$.

(a) How many bands per unit cell do you expect to find?

(b) Write down an expression for a tight-binding wavefunction for this system.

(c) Assuming that the atomic orbitals are orthonormal, write down the Slater–Koster expressions for the types of bonds you expect in this system. Draw a sketch of each type of bond. Choose a coordinate system on one atom in the unit cell and use it consistently.

(d) Write out explicitly all terms in the secular determinant.

(e) Construct the first Brillouin zone of the simple cubic real space lattice. Label the special points and find the symmetry operations associated with them.

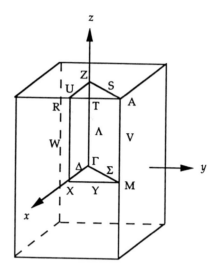

Fig. 15.15.   The Brillouin zone for the primitive tetragonal lattice.

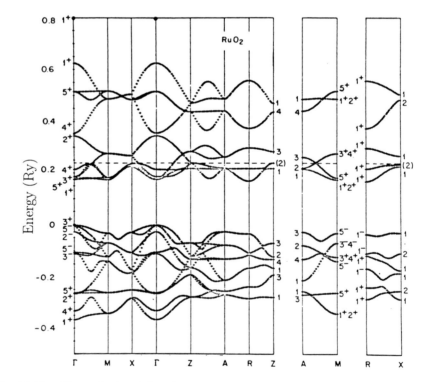

Fig. 15.16.   The calculated band structure for RuO_2. From L.F. Mattheiss, *Phys. Rev. B* **13**, 2433 (1976). Used with permission.

(f) Assuming first that the *s* and *d* orbital interactions can be decoupled from one another, solve for the *s* bands and then the *d* bands separately. You need also assume that $m = 1$ and $n = 5$ to get the full solution. Make sketches for the band structure around a convenient circuit in the Brillouin zone. If you cannot solve the equations explicitly, make a qualitative sketch. Be certain that you justify the slope of each band at all special points and that the widths of the various bands are in the correct proportions. Now assume that we switch on the *s*–*d* interactions. How does the band structure change?

(g) From your full band structure, *sketch* (do not calculate) a qualitative DOS.

(h) If the crystal structure were to elongate along the *z* axis, how would you expect bonding to be affected as a function of N_d? Does this give you any insight into bonding in the high-T_c superconductor $YBa_2Cu_3O_7$?

15.2 Derive Eqs. 15.12, 15.13, 15.14, and 15.15. Make schematic plots of R_0, $\mathscr{E}_{coh}$, W, and B, as a function of N_d ($0 \leq N_d \leq 10$) and explain the features that you see in each plot.

16

More sophisticated electronic structure calculations

16.1 Introduction

Up to this point we have studied a wide variety of materials by exploiting a few basic principles. Specifically, by examining the constituent atoms' valence electron configurations, we have been able to write down an appropriate set of basis functions for the solid. Whether these sets are comprised of atomic orbitals, plane waves, or sums of plane waves, they can be inserted into the Schrödinger equation and the eigenvalues of the system can be found. We have also seen how translational symmetry can be used to significant advantage, both by reducing the number of independent degrees of freedom in a system and by introducing reciprocal space. The band structure of a material gives valuable insight into its electronic, optical, and magnetic properties. Admittedly, interpretation of what have been named 'spaghetti diagrams' is often opaque and does not necessarily lead to an understanding of bonding in the material. Still, the density of states combined with the moments theorem allows us to understand the relationship between the local atomic environment and bonding in systems, even when they lack translational symmetry.

But we have made a number of drastic simplifying assumptions along the way. In particular, we have, save for the brief discussion of transition metals in Ch. 15, ignored the Coulomb interaction between electrons. Furthermore, we have ignored any *exchange* effects between electrons which arise because of their spin, except to recognize that the Fermi distribution function is the proper function for populating levels at $T > 0$ K. The reason for this *independent electron approximation* is obvious enough. In any reasonable size of material there are on the order of 10^{23} electrons. Imagine trying to solve for the position and velocity of each of these electrons when they each have a kinetic energy contribution, a repulsive Coulomb interaction with other electrons in the system, an attractive interaction with the nuclei, and an interaction that arises from the Pauli exclusion principle to *correlate* the electrons' motion. Clearly this is an intractable problem.

A complete Hamiltonian includes *all* of the above interactions,

$$\mathscr{H} = \mathscr{H}_{KE} + \mathscr{H}_C + \mathscr{H}_{ionic} + \mathscr{H}_{ex-corr} \tag{16.1}$$

where $\mathscr{H}_{KE}$ is the usual kinetic energy contribution,

$$\mathscr{H}_{KE} = -\frac{\hbar^2}{2m}\nabla^2 \tag{16.2}$$

$\mathscr{H}_C$ is the electron–electron Coulomb repulsion, $\mathscr{H}_{ionic}$ is the electron–ion interaction, and $\mathscr{H}_{ex-corr}$ is the exchange–correlation interaction. The final three terms are the potential energy contributions to the Hamiltonian. One of the problems that we have faced continually is the appropriate choice for the functional form of the potential energy. Breaking it down into separate contributions helps us to focus on the various interactions and, indeed, different theories have been developed for each term.

These theories should begin from elementary ingredients such as properties of the nuclei, the electrons, and their interactions. Such theories are known as *ab initio* or first-principles theories. Ideally, no approximations are made. For practical calculations, however, a number of approximations are made, although these are rigorously tested. The tight-binding theory, with the interaction parameters $(ss\sigma), (sp\sigma)$, *etc.*, is a *semi-empirical* theory in that it relies on input of experimental data to adjust the parameters.

The final feature that we point out about our calculations so far is the fact that they are not *self-consistent*. By this we mean that we choose what we believe is an appropriate potential and solve for the eigenvalues and eigenfunctions. However, a moment's reflection on the various contributions to the complete Hamiltonian shows that they *depend on the wavefunctions!* How are we supposed to solve for the wavefunctions when we need them as input to the calculation in the first place? This dilemma points to the need for self-consistent calculations: an initial guess for the potential is made; the Schrödinger equation is solved and these wavefunctions are used to construct a *new* potential. The new potential is inserted into the Schrödinger equation and solved for a new set of eigenvalues and wavefunctions. The process is repeated until the output potential differs from that of the previous step by less than some desired accuracy. In other words, the calculation is self-consistent. This strategy is outlined in Fig. 16.1.

In the remaining sections of this chapter we will discuss the various contributions to the complete Hamiltonian. We conclude with a survey of

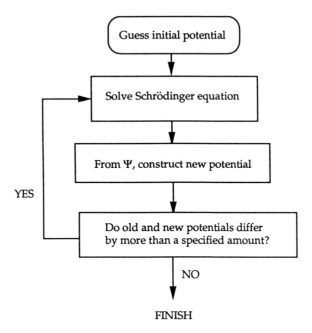

Fig. 16.1. Outline of the approach to self-consistent electronic structure calculations.

the types of problems that have been studied with these more sophisticated electronic structure calculations.

16.2 Electron–electron interactions

Great strides in understanding the properties of the interacting electron gas were made in the early 1960s. At that time, Hohenberg and Kohn showed that the ground state properties of an interacting electron gas can be calculated from the electron density, *independent* of the nature of the external potential. Their ideas followed earlier work from the 1920s by Thomas and Fermi.

The Hohenberg–Kohn theorem† may be summarized as follows:

Theorem 16.1 *The ground-state energy of a system of interacting electrons is a unique functional of the charge density. It is assumed that the ground state is nondegenerate. This functional has its minimum value for the correct ground-state energy for variations in the charge density when the number of electrons is kept fixed.*

† P. Hohenberg and W. Kohn, *Phys. Rev.* **136**, B864 (1964).

A *functional* is a function of a function. For example, the charge density, n, is a function of position, $\vec{r}$. Thus, the functional that gives the correct ground state energy is denoted by $F[n(\vec{r}\,)]$, which is a function of a function. While the Hohenberg–Kohn theorem provided justification for focusing on the electron density, it gave no insight into the nature of the functional $F[n(\vec{r}\,)]$. The Hohenberg–Kohn theorem is only an *existence* theorem. Because of the central importance of the electron density, this theory of the electron gas came to be known as Density Functional Theory.

In the following year, Kohn and Sham† provided a procedure by which we can actually solve for the correct $n(\vec{r}\,)$. They considered a system of N *noninteracting* electrons in an arbitrary external potential, $v_{\text{ext}}(\vec{r}\,)$. Let the ground state density of this system be denoted by $n(\vec{r}\,)$. The ground-state-energy functional is written as:

$$E[n] = \int v_{\text{ext}}(\vec{r}\,)\, n(\vec{r}\,)\, d\vec{r} + F[n] \qquad (16.3)$$

where $v_{\text{ext}}(\vec{r}\,)$ is the external potential and the functional $F[n]$ includes all the kinetic energy and electron–electron interaction terms. Most frequently, the Coulomb interaction is separated out from the functional $F[n]$,

$$F[n] = \frac{e^2}{8\pi\epsilon_0} \int \int \frac{n(\vec{r}\,)n(\vec{r}'\,)}{|\vec{r} - \vec{r}'|}\, d\vec{r}\, d\vec{r}' + G[n] \qquad (16.4)$$

where e is the (negative) electron charge, ϵ_0 is the permittivity of free space, and the Coulomb interaction can be thought of as the interaction between two charge *densities*, separated by a distance $|\vec{r} - \vec{r}'|$, rather than the more familiar interaction between two *point* charges. The new functional $G[n]$ remains of unknown form.

Using the calculus of variations, the ground state energy is found by minimizing the energy in Eq. 16.3 with respect to variations in the electron density for which particle number is conserved,

$$\int n(\vec{r}\,)\, d\vec{r} = N \qquad (16.5)$$

The calculus of variations is an exquisite and powerful theory that is used extensively to search through the space of *functions* which extremize a quantity subject to some constraint(s). One familiar example is to find the equation for the line of shortest distance between two points fixed on the surface of a sphere. Anyone who has ever taken an intercontinental flight knows that this line is a great circle. The variational equation for the ground

† W. Kohn and L.J. Sham, *Phys. Rev.* **140**, A1133 (1965).

state energy subject to the constraint Eq. 16.5 is

$$\delta\left\{E[n] - \mu \int n(\vec{r}\,)\,d\vec{r}\right\} = 0 \tag{16.6}$$

where δ means take the variational derivative with respect to $n(\vec{r}\,)$ and μ is a Lagrange multiplier which assures that our constraint is obeyed. Inserting Eqs. 16.3 and 16.4 into Eq. 16.6,

$$v_{\text{ext}}(\vec{r}\,) + \frac{e^2}{4\pi\epsilon_0} \int \frac{n(\vec{r}\,)}{|\vec{r} - \vec{r}'|}\,d\vec{r} + \frac{\delta G[n]}{\delta n} = \mu \tag{16.7}$$

where the last term on the left-hand side is the variational derivative of the functional $G[n]$ with respect to $n(\vec{r}\,)$.

For the Kohn–Sham system of *noninteracting* electrons, Eq. 16.7 reduces to

$$v_{\text{ext,n.i.}}(\vec{r}\,) + \frac{\delta T_{\text{n.i.}}[n]}{\delta n(\vec{r}\,)} = \mu_{\text{n.i.}} \tag{16.8}$$

where $T_{\text{n.i.}}[n]$ is the kinetic energy functional and the subscripts n.i. everywhere mean 'noninteracting'. $T[n]$ is generally of unknown form, but there is an alternative way of finding $n(\vec{r}\,)$. The wavefunction for the ground state of the noninteracting system can be written down readily! Briefly, this many-particle wavefunction is a product of one-electron wavefunctions that is 'antisymmetrized', to reflect the spin of the electrons.† Each one-electron wavefunction, $\phi_i(\vec{r}\,)$, obeys the Schrödinger equation

$$\left\{-\frac{\hbar^2}{2m}\nabla^2 + v_{\text{ext,n.i.}}(\vec{r}\,)\right\}\phi_i(\vec{r}\,) = \mathscr{E}_i\phi_i(\vec{r}\,) \tag{16.9}$$

for $i = 1, ..., N$. The density of this system is

$$n(\vec{r}\,) = \sum_{i=1}^{N}|\phi_i(\vec{r}\,)|^2 \tag{16.10}$$

The important contribution of Kohn and Sham was to generalize this to the case of *interacting* electrons. The functional $G[n]$ from Eq. 16.4 is split up into two terms,

$$G[n] = T_{\text{n.i.}}[n] + E_{xc}[n] \tag{16.11}$$

where $T_{\text{n.i.}}[n]$ is the kinetic energy functional for the system of noninteracting electrons (Eq. 16.8) and $E_{xc}[n]$ is called the exchange and correlation energy of the interacting system of density $n(\vec{r}\,)$.

---

† An antisymmetrized wavefunction will change sign under the interchange of any two particles.

The variational equation for the electron density, Eq.16.7, now is

$$v_{\text{ext}}(\vec{r}\,) + \frac{e^2}{4\pi\epsilon_0} \int \frac{n(\vec{r}^{\,\prime}\,)}{|\vec{r} - \vec{r}^{\,\prime}|}\, d\vec{r}^{\,\prime} + \frac{\delta E_{xc}[n]}{\delta n(\vec{r}\,)} + \frac{\delta T_{\text{n.i.}}[n]}{\delta n(\vec{r}\,)} = \mu \qquad (16.12)$$

We can define an effective potential by

$$v_{\text{eff}}(\vec{r}\,) \equiv v_{\text{ext}}(\vec{r}\,) + \frac{e^2}{4\pi\epsilon_0} \int \frac{n(\vec{r}^{\,\prime})}{|\vec{r} - \vec{r}^{\,\prime}|}\, d\vec{r}^{\,\prime} + \frac{\delta E_{xc}[n]}{\delta n(\vec{r}\,)} \qquad (16.13)$$

Therefore, analogous to the noninteracting case, the correct ground state energy density of the interacting system is obtained by solving the 'Kohn–Sham' equations,

$$\left\{ -\frac{\hbar^2}{2m}\nabla^2 + v_{\text{eff}}[n] \right\} \psi_i(\vec{r}\,) = \mathscr{E}_i \psi_i(\vec{r}\,) \qquad (16.14)$$

where the sum runs over $i = 1, ..., N$, and

$$n(\vec{r}\,) = \sum_{i=1}^{N} |\psi_i(\vec{r}\,)|^2 \qquad (16.15)$$

The functions $\psi_i(\vec{r}\,)$ are one-particle wavefunctions and Eqs. 16.14 and 16.15 must be solved self-consistently because v_{eff} depends on $n(\vec{r}\,)$. Note that the Kohn–Sham equations have a form functionally identical with that of our ordinary Schrödinger equation. The ground state energy of the interacting electron system is thus given by

$$\begin{aligned} E[n] = {}& \sum_{i=1}^{N} \int \psi_i^*(\vec{r}\,) \left(-\frac{\hbar^2}{2m}\nabla^2 \right) \psi_i(\vec{r}\,)\, d\vec{r} \\ &+ \int v_{\text{ext}}(\vec{r}\,)\, n(\vec{r}\,)\, d\vec{r} + \frac{e^2}{8\pi\epsilon_0} \int\int \frac{n(\vec{r}\,)\, n(\vec{r}^{\,\prime}\,)}{|\vec{r} - \vec{r}^{\,\prime}|}\, d\vec{r}\, d\vec{r}^{\,\prime} \\ &+ E_{xc}[n] \end{aligned} \qquad (16.16)$$

But what is this exchange and correlation energy? The exchange interaction arises from the Pauli exclusion principle, which requires electrons with parallel spins to avoid one another. Clearly this is a repulsive interaction and raises the ground state energy of the system. The correlation energy actually has no rigorous physical interpretation, but rather is a correction term to compensate for a subtle approximation made. We will not delve into the technical details any further here.

In order to solve the Kohn–Sham equations, however, some explicit functional form is required for the exchange–correlation functional. One very popular approximation is the *Local Density Approximation,*

$$E_{xc}[n] = \int \mathscr{E}_{xc}(n(\vec{r}\,))\,n(\vec{r}\,)\,\mathrm{d}\vec{r} \tag{16.17}$$

$\mathscr{E}_{xc}(n)$ is the exchange and correlation energy for an interacting electron gas of *uniform* density. Equation 16.17 takes its name because the energy functional at a point $\vec{r}$ depends only on the electron density at $\vec{r}$. This approximation is clearly valid for systems in which the electron density is uniform and is considered an adequate approximation in systems with slowly varying $n(\vec{r}\,)$. There are several models for which $\mathscr{E}_{xc}(n)$ has been calculated.

16.3 Electron–ion interactions

Solution of the Kohn–Sham equations requires, in addition to understanding electron–electron interactions, some knowledge of the appropriate form for the electron–ion interaction. We have pointed out before, and it is certainly borne out by observation, that the properties of solids (and atoms and molecules as well) are determined primarily by the valence atomic electrons. We would like to be able to formalize this observation when developing a theory of electron–ion interactions. Although Fermi was the first to broach such ideas in the 1930s, Phillips and Kleinman[†] were the first to develop the 'pseudopotential' theory in 1959. We will present the original arguments of Phillips and Kleinman and show why it is possible to ignore the core electrons.

Let $\psi_{\vec{k}}^{(n)}(\vec{r}\,)$ be the wavefunction for a state of wavevector $\vec{k}$ and band index n. For a system in which the bands are broad (strong overlap), the wavefunction must look like a plane wave far away from any nucleus. Closer to the nucleus, however, the wavefunction will vary rapidly (have many nodes) and will be better represented by an atomic orbital $\phi_i(\vec{r}\,)$, where the index j runs over all core states. The valence electron wavefunction can thus be written as a smooth *pseudowavefunction* $f_{\vec{k}}^{(n)}(\vec{r}\,)$ corrected to be orthogonal to all the core states, $\phi_i(\vec{r}\,)$,

$$|\psi_{\vec{k}}^{(n)}\rangle = |f_{\vec{k}}^{(n)}\rangle - \sum_j |\phi_j\rangle\langle\phi_j|f_{\vec{k}}^{(n)}\rangle \tag{16.18}$$

Exercise 16.1   Show that Eq. 16.18 is orthogonal to the core states $|\phi_j\rangle$.

Substituting Eq. 16.18 into the Schrödinger equation and assuming that

† J.C. Phillips and L. Kleinman, *Phys. Rev.* **116**, 287 (1959).

the core wavefunctions are eigenfunctions of $\mathcal{H}$, we find

$$\mathcal{H}|f_{\vec{k}}^{(n)}\rangle + \sum_j (\mathcal{E} - \mathcal{E}_j)|\phi_j\rangle\langle\phi_j|f_{\vec{k}}^{(n)}\rangle = \mathcal{E}|f_{\vec{k}}^{(n)}\rangle \qquad (16.19)$$

or, explicitly dividing the Hamiltonian into its kinetic energy, $\mathcal{T}$, and potential energy, $\mathcal{V}$, contributions, yields

$$\mathcal{T}|f_{\vec{k}}^{(n)}\rangle + \mathcal{V}_{ps}|f_{\vec{k}}^{(n)}\rangle = \mathcal{E}|f_{\vec{k}}^{(n)}\rangle \qquad (16.20)$$

where $\mathcal{V}_{ps}$ is the *pseudopotential*,

$$\mathcal{V}_{ps} \equiv \mathcal{V} + \sum_j (\mathcal{E} - \mathcal{E}_j)|\phi_j\rangle\langle\phi_j| \qquad (16.21)$$

Exercise 16.2 Derive Eqs. 16.19 and 16.20.

Note that the pseudopotential of Eq. 16.21 is *weaker* than the original potential, $\mathcal{V}$, because $\mathcal{E}_j < \mathcal{E}$ and, hence, the second term is repulsive.

Exercise 16.3 Why is $\mathcal{E}_j < \mathcal{E}$?

A physical way to think about the weakness of the pseudopotential is to recognize that if the valence wavefunction has many nodes close to the nucleus, it has a high kinetic energy. If the electron's kinetic energy is large, then it spends little time in the vicinity of the nucleus. The effective pseudowavefunction, $|f_{\vec{k}}^{(n)}\rangle$, can thus be depicted as a function that is smoothly varying for all distances, as shown in Fig. 16.2.

16.4 Some applications

Ab initio electronic structure calculations of the kind described above have been performed on a wide variety of systems since the 1970s. Recall the list of materials characterized by structure that we presented in Ch. 1; these sophisticated techniques have been used to study properties of virtually every class of materials listed there. Calculated band structures have been compared with experiment, and optical, transport, and mechanical properties have been investigated.

There are serious drawbacks to these techniques, however. For one, only a small number of atoms can be treated by these sophisticated methods. Therefore, studies of large-scale defects such as grain boundaries are limited to cases with short periods. At present, it is not routinely possible to study systems with more than approximately 100 atoms. Furthermore, systems in which the exchange and correlation interaction is large are difficult to study; these include the 'heavy fermion' systems. Finally, by virtue of the fact that

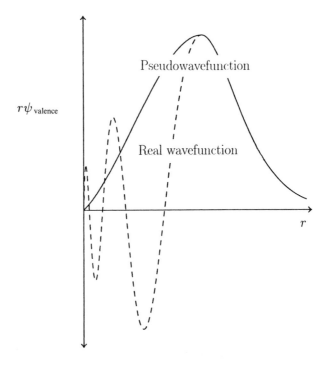

Fig. 16.2. Schematic diagram of the real wavefunction with nodes and the pseudo-wavefunction without nodes in the core region. Note that beyond some radius the two wavefunctions must be identical.

the Kohn–Sham equations are most often solved in reciprocal space (which entails expanding all quantities of interest in Fourier series), it is cumbersome to describe accurately the potential of some elements. These elements include the highly electronegative F and Cl and the first-row elements C and O. The difficulty arises from the number of wiggles in the potential close to the nucleus; short real-space distances become large reciprocal-space distances and, hence, require many plane waves for accurate series. Calculations involving these elements become very computer-intensive.

One of the biggest developments in recent years, however, is the calculation of the *total energy*. The total energy of a system of electrons and ions is given by

$$\mathscr{E}_T = \mathscr{E}_{el} + \mathscr{E}_{e-ion} + \mathscr{E}_{ion-ion} \tag{16.22}$$

where $\mathscr{E}_{el}$ is the electron total energy, $\mathscr{E}_{e-ion}$ is the electron–ion interaction energy, and $\mathscr{E}_{ion-ion}$ is the ion–ion interaction energy.† In addition to solving

† J. Ihm, A. Zunger, and M.L. Cohen, *J. Phys. C* **12**, 4409 (1979).

the Schrödinger equation for the electronic energy levels, the other terms in Eq. 16.22 are found. Knowledge of the total energy of a solid affords a wide array of possible calculations. For example, total energy calculations have been used to assess the relative stability of crystal structures, to examine changes in crystal structure with applied external pressure, to construct rudimentary phase diagrams, to determine formation and migration energies for defects in semiconductors and metals, and to elucidate the structure of defects. The interested reader is referred to the literature for further examples.

Despite the enormous success enjoyed by *ab initio* electronic structure calculations, it has been recognized that they are not a panacea for all bonding questions in solids. Indeed, there is a trend now toward developing realistic potentials in which we can bypass the complicated band structure calculations and still arrive at the total energy. There is a long history to these developments, beginning in the 1920s with the advent of the Lennard-Jones empirical potential. Such *pair* potentials, which only admit interaction between any two pairs of atoms, have been relatively successful in modeling structural properties of metallic systems. Metallic systems work well because they are often, as we have seen, close-packed. However, covalent or ionic systems are more challenging. Using some of the considerations that we have outlined in this book, potentials that model interactions in open systems† have been developed. An outstanding problem that remains is how to treat close-packed systems that are in contact with open systems, such as one might find in a metal–ceramic composite.

---

† That is, not close-packed.

Appendix 1

Atomic units

Atomic units, which are also known as Hartree units, are defined as follows. The unit of mass is the electron mass, m_e, as opposed to the gram; the unit of charge is the proton charge, e, as opposed to the statcoulomb or the coulomb; the unit of angular momentum is $\hbar$ rather than g cm^2/s.

The ground state energy of the hydrogen atom is given by $-\frac{1}{2}Z^2(e^2/a_0)$, where the Bohr radius is $a_0 = \hbar^2/m_e e^2$ and has a numerical value of 1 in atomic units.

The atomic unit of energy, e^2/a_0, is known as the hartree:

$$1 \text{ hartree} = \frac{e^2}{a_0} = 27.212 \text{ eV} \tag{A1.1}$$

Note that the ground state energy of the hydrogen atom is $-\frac{1}{2}$ hartree, provided nuclear motion is neglected. Unfortunately, the hartree seems to be a unit more popular in Europe. The unit of choice in the US (and the rest of North America to the best of my knowledge) is the rydberg (Ry):

$$1 \text{ rydberg} = \frac{e^2}{2a_0} = \frac{1}{2} \text{ hartree} \tag{A1.2}$$

The atomic unit of length is the bohr:

$$1 \text{ bohr} = a_0 = 0.529\,18 \text{ Å} \tag{A1.3}$$

Recall that a_0 is the most probable radius at which to find an electron in the ground state of a hydrogen atom.

Appendix 2

Stereographic projections of some point groups

The point groups are divided up into the crystal system to which they belong. The technique for generating stereographic projections as well as the symbols for various symmetry elements are discussed in detail in Ch. 2.

Triclinic	Monoclinic (1)	Tetragonal
C_1 (1)	C_2 (2)	C_4 (4)
——	C_{1h} (m)	S_4 ($\bar{4}$)
S_1 ($\bar{1}$)	C_{2h} (2/m)	C_{4h} (4/m)

Monoclinic (2)	Orthorhombic	
C_2 (2)	D_2 (222)	D_4 (422)
C_{1h} (m)	C_{2v} (mm2)	C_{4v} (4mm)
——	——	D_{2d} ($\bar{4}$2m)
C_{2h} (2/m)	D_{2h} (mmm)	D_{4h} (4/mmm)

Trigonal	Hexagonal	Cubic
C_3 (3)	C_6 (6)	T (23)
—	C_{3h} ($\bar{6}$)	—
S_3 ($\bar{3}$)	C_{6h} (6/m)	T_h (m3)
D_3 (32)	D_6 (622)	O (432)
C_{3v} (3m)	C_{6v} (6mm)	—
—	D_{3h} ($\bar{6}$m2)	T_d ($\bar{4}$3m)
D_{3d} ($\bar{3}$m)	D_{6h} (6/mmm)	O_h (m3m)

Appendix 3
Character tables

The character tables for common point groups are given in the following pages. Each section contains groups of similar structure, so that, for example, all pure rotational groups are listed in the section labeled 'C_n Groups'. All tables are constructed in the same way. The leftmost column contains the group symbol in the first row, followed by the irreducible representation symbols. The classes of symmetry operations are listed across the top row, while the actual characters comprise the bulk of the table. At the rightmost of each table are listed the functions that generate the irreducible representations in the respective row. More details about construction of these tables are found in Ch. 3.

Non-axial Groups

C_1	E
A	1

S_2	E	i		
A_g	1	1	R_x, R_y, R_z	x^2, y^2, z^2
				xy, xz, yz
A_u	1	-1	x, y, z	

C_{1h}	E	σ_h		
A'	1	1	x, y, R_z	x^2, y^2, z^2, xy
A''	1	-1	z, R_x, R_y	xy, xz

C_n Groups

C_2	E	C_2		
A	1	1	z, R_z	x^2, y^2, z^2, xy
B	1	-1	x, y, R_z, R_y	yz, xz

C_3	E	C_3	C_3^2			$\epsilon = \exp(2\pi\iota/3)$
A	1	1	1	z, R_z		$x^2 + y^2, z^2$
E	1	ϵ	ϵ^*	$x + \iota y, R_x + \iota R_y$		$(x^2 - y^2, xy)$
	1	ϵ^*	ϵ	$x - \iota y, R_x - \iota R_y$		(yz, xz)

C_4	E	C_4	C_2	C_4^3		
A	1	1	1	1	z, R_z	$x^2 + y^2, z^2$
B	1	-1	1	-1		$x^2 - y^2, xy$
E	1	ι	-1	$-\iota$	$x + \iota y, R_x + \iota R_y$	(yz, xz)
	1	$-\iota$	-1	ι	$x - \iota y, R_x - \iota R_y$	

C_5	E	C_5	C_5^2	C_5^3	C_5^4			$\epsilon = \exp(2\pi\iota/5)$
A	1	1	1	1	1	z, R_z		$x^2 + y^2, z^2$
E_1	1	ϵ	ϵ^2	ϵ^{2*}	ϵ^*	$x + \iota y, R_x + \iota R_y$		(yz, xz)
	1	ϵ^*	ϵ^{2*}	ϵ^2	ϵ	$x - \iota y, R_x - \iota R_y$		
E_2	1	ϵ^2	ϵ^*	ϵ	ϵ^{2*}			$(x^2 - y^2, xy)$
	1	ϵ^{2*}	ϵ	ϵ^*	ϵ^2			

C_6	E	C_6	C_3	C_2	C_3^2	C_6^5		$\epsilon = \exp(2\pi i/6)$
A	1	1	1	1	1	1	z, R_z	x^2+y^2, z^2
B	1	-1	1	-1	1	-1		
E_1	1	ϵ	$-\epsilon^*$	-1	$-\epsilon$	ϵ^*	$x+iy, R_x+iR_y$	(xz, yz)
	1	ϵ^*	$-\epsilon$	-1	$-\epsilon^*$	ϵ	$x-iy, R_x+iR_y$	
E_2	1	$-\epsilon^*$	$-\epsilon$	1	$-\epsilon^*$	$-\epsilon$		(x^2-y^2, xy)
	1	$-\epsilon$	$-\epsilon^*$	1	$-\epsilon$	$-\epsilon^*$		

C_7	E	C_7	C_7^2	C_7^3	C_7^4	C_7^5	C_7^6		$\epsilon = (2\pi i/7)$
A	1	1	1	1	1	1	1	z, R_z	x^2+y^2, z^2
E_1	1	ϵ	ϵ^2	ϵ^3	ϵ^{3*}	ϵ^{2*}	ϵ^*	$x+iy, R_x+iR_y$	(xz, yz)
	1	ϵ^*	ϵ^{2*}	ϵ^{3*}	ϵ^3	ϵ^2	ϵ	$x-iy, R_x+iR_y$	
E_2	1	ϵ^2	ϵ^{3*}	ϵ^*	ϵ	ϵ^3	ϵ^{2*}		(x^2-y^2, xy)
	1	ϵ^{2*}	ϵ^3	ϵ	ϵ^*	ϵ^{3*}	ϵ^2		
E_3	1	ϵ^3	ϵ^*	ϵ^2	ϵ^{2*}	ϵ	ϵ^{3*}		
	1	ϵ^{3*}	ϵ	ϵ^{2*}	ϵ^2	ϵ^*	ϵ^3		

C_8	E	C_8	C_4	C_8^3	C_2	C_8^5	C_4^3	C_8^7		$\epsilon = (2\pi i/8)$
A	1	1	1	1	1	1	1	1	z, R_z	x^2+y^2, z^2
B	1	-1	1	-1	1	-1	1	-1		
E_1	1	ϵ	i	$-\epsilon^*$	-1	$-\epsilon$	$-i$	ϵ^*	$x+iy, R_x+iR_y$	(xz, yz)
	1	ϵ^*	$-i$	$-\epsilon$	-1	$-\epsilon^*$	i	ϵ	$x-iy, R_x-iR_y$	
E_2	1	i	-1	$-i$	1	i	-1	$-i$		(x^2-y^2, xy)
	1	$-i$	-1	i	1	$-i$	-1	i		
E_3	1	$-\epsilon^*$	i	ϵ	-1	ϵ^*	$-i$	$-\epsilon$		
	1	$-\epsilon$	$-i$	ϵ^*	-1	ϵ	i	$-\epsilon^*$		

C_{nv} Groups

C_{2v}	E	C_2	σ_v	σ_v'		
A_1	1	1	1	1	z	x^2, y^2, z^2
A_2	1	1	-1	-1	R_z	xy
B_1	1	-1	1	-1	x, R_y	xz
B_2	1	-1	-1	1	y, R_x	yz

C_{3v}	E	$2C_3$	$3\sigma_v$		
A_1	1	1	1	z	$x^2 + y^2, z^2$
A_2	1	1	-1	R_z	
E	2	-1	0	$(x, y)(R_x, R_y)$	$(x^2 - y^2, xy)(xz, yz)$

C_{4v}	E	$2C_4$	C_2	$2\sigma_v$	$2\sigma_d$		
A_1	1	1	1	1	1	z	$x^2 + y^2, z^2$
A_2	1	1	1	-1	-1	R_z	
B_1	1	-1	1	1	-1		$x^2 - y^2$
B_2	1	-1	1	-1	1		xy
E	2	0	-2	0	0	$(x, y)(R_x, R_y)$	(xz, yx)

C_{5v}	E	$2C_5$	$2C_5^2$	$5\sigma_v$	$\alpha = 2\cos 72°$	$\beta = 2\cos 144°$
A_1	1	1	1	1	z	$x^2 + y^2, z^2$
A_2	1	1	1	-1	R_z	
E_1	2	α	β	0	$(x, y)(R_x, R_y)$	(xz, yz)
E_2	2	β	α	0		$(x^2 - y^2, xy)$

C_{6v}	E	$2C_6$	$2C_3$	C_2	$3\sigma_v$	$3\sigma_d$		
A_1	1	1	1	1	1	1	z	$x^2 + y^2, z^2$
A_2	1	1	1	1	-1	-1	R_z	
B_1	1	-1	1	-1	1	-1		
B_2	1	-1	1	-1	-1	1		
E_1	2	1	-1	-2	0	0	$(x, y)(R_x, R_y)$	(xz, yz)
E_2	2	-1	-1	2	0	0		$(x^2 - y^2, xy)$

C_{nh} Groups

C_{2h}	E	C_2	i	σ_h		
A_g	1	1	1	1	R_z	x^2, y^2, z^2, xy
B_g	1	-1	1	-1	R_x, R_y	xz, yz
A_u	1	1	-1	-1	z	
B_u	1	-1	-1	1	x, y	

C_{3h}	E	C_3	C_3^2	σ_h	S_3	S_3^5		$\epsilon = \exp(2\pi i/3)$
A'	1	1	1	1	1	1	R_z	x^2+y^2, z^2
E'	$\left\{\begin{matrix}1\\1\end{matrix}\right.$	$\begin{matrix}\epsilon\\\epsilon^*\end{matrix}$	$\begin{matrix}\epsilon^*\\\epsilon\end{matrix}$	$\begin{matrix}1\\1\end{matrix}$	$\begin{matrix}\epsilon\\\epsilon^*\end{matrix}$	$\begin{matrix}\epsilon^*\\\epsilon\end{matrix}$	$\begin{matrix}x+iy\\x-iy\end{matrix}$	(x^2-y^2, xy)
A''	1	1	1	-1	-1	-1	z	
E''	$\left\{\begin{matrix}1\\1\end{matrix}\right.$	$\begin{matrix}\epsilon\\\epsilon^*\end{matrix}$	$\begin{matrix}\epsilon^*\\\epsilon\end{matrix}$	$\begin{matrix}-1\\-1\end{matrix}$	$\begin{matrix}-\epsilon\\-\epsilon^*\end{matrix}$	$\begin{matrix}-\epsilon^*\\-\epsilon\end{matrix}$	$\begin{matrix}R_x+iR_y\\R_x-iR_y\end{matrix}$	(xz, yz)

C_{4h}	E	C_4	C_2	C_4^3	i	S_4^3	σ_h	S_4		
A_g	1	1	1	1	1	1	1	1	R_z	x^2+y^2, z^2
B_g	1	-1	1	-1	1	-1	1	-1		x^2-y^2, xy
E_g	$\left\{\begin{matrix}1\\1\end{matrix}\right.$	$\begin{matrix}i\\-i\end{matrix}$	$\begin{matrix}-1\\-1\end{matrix}$	$\begin{matrix}-i\\i\end{matrix}$	$\begin{matrix}1\\1\end{matrix}$	$\begin{matrix}i\\-i\end{matrix}$	$\begin{matrix}-1\\-1\end{matrix}$	$\begin{matrix}-i\\i\end{matrix}$	$\begin{matrix}R_x+iR_y\\R_x-iR_y\end{matrix}$	(xz, yz)
A_u	1	1	1	1	-1	-1	-1	-1	z	
B_u	1	-1	1	-1	-1	1	-1	1		
E_u	$\left\{\begin{matrix}1\\1\end{matrix}\right.$	$\begin{matrix}i\\-i\end{matrix}$	$\begin{matrix}-1\\-1\end{matrix}$	$\begin{matrix}-i\\i\end{matrix}$	$\begin{matrix}-1\\-1\end{matrix}$	$\begin{matrix}-i\\i\end{matrix}$	$\begin{matrix}1\\1\end{matrix}$	$\begin{matrix}i\\-i\end{matrix}$	$\begin{matrix}x+iy\\x-iy\end{matrix}$	

D_n Groups

D_2	E	C_2	$C_2(y)$	$C_2(x)$		
A	1	1	1	1		x^2, y^2, z^2
B_1	1	1	-1	-1	z, R_z	xy
B_2	1	-1	1	-1	y, R_y	xz
B_3	1	-1	-1	1	x, R_x	yz

D_3	E	$2C_3$	$3C_2'$		
A_1	1	1	1		x^2+y^2, z^2
A_2	1	1	-1	z, R_z	
E	2	-1	0	$(x,y)(R_x, R_y)$	$(x^2-y^2, xy)(xz, yz)$

D_4	E	$2C_4$	$C_2(\equiv C_4^2)$	$2C_2'$	$2C_2''$		
A_1	1	1	1	1	1		x^2+y^2, z^2
A_2	1	1	1	-1	-1	z, R_z	
B_1	1	-1	1	1	-1		x^2-y^2
B_2	1	-1	1	-1	1		xy
E	2	0	-2	0	0	$(x,y)(R_x, R_y)$	(xz, yz)

D_5	E	$2C_5$	$2C_5^2$	$5C_2'$	$\alpha = 2\cos 72°$	$\beta = 2\cos 144°$	
A_1	1	1	1	1			x^2+y^2, z^2
A_2	1	1	1	-1	z, R_z		
E_1	2	α	β	0	$(x,y)(R_x, R_y)$	(xz, yz)	
E_2	2	β	α	0		(x^2-y^2, xy)	

D$_{nd}$ Groups

D_{2d}	E	$2S_4$	C_2	$2C_2'$	$2\sigma_d$		
A_1	1	1	1	1	1		x^2+y^2, z^2
A_2	1	1	1	-1	-1	R_z	
B_1	1	-1	1	1	-1		x^2-y^2
B_2	1	-1	1	-1	1	z	xy
E	2	0	-2	0	0	$(x,y)(R_x,R_y)$	(xz,yz)

D_{3d}	E	$2C_3$	$3C_2'$	i	$2S_6$	$3\sigma_d$		
A_{1g}	1	1	1	1	1	1		x^2+y^2, z^2
A_{2g}	1	1	-1	1	1	-1	R_z	
E_g	2	-1	0	2	-1	0	(R_x,R_y)	$(x^2-y^2,xy)(xz,yz)$
A_{1u}	1	1	1	-1	-1	-1		
A_{2u}	1	1	-1	-1	-1	1	z	
E_u	2	-1	0	-2	1	0	(x,y)	

D_{4d}	E	$2S_8$	$2C_4$	$2S_8^3$	C_2	$4C_2'$	$4\sigma_d$		
A_1	1	1	1	1	1	1	1		x^2+y^2, z^2
A_2	1	1	1	1	1	-1	-1	R_z	
B_1	1	-1	1	-1	1	1	-1		
B_2	1	-1	1	-1	1	-1	1	z	
E_1	2	$\sqrt{2}$	0	$-\sqrt{2}$	-2	0	0	(x,y)	
E_2	2	0	-2	0	2	0	0		(x^2-y^2,xy)
E_3	2	$-\sqrt{2}$	0	$\sqrt{2}$	-2	0	0	(R_x,R_y)	(xz,yz)

D_{5d}	E	$2C_5$	$2C_5^2$	$5C_2'$	i	$2S_{10}^3$	$2S_{10}$	$5\sigma_d$	$\alpha = 2\cos 70°$	$\beta = 2\cos 144°$
A_{1g}	1	1	1	1	1	1	1	1		x^2+y^2, z^2
A_{2g}	1	1	1	-1	1	1	1	-1	R_z	
E_{1g}	2	α	β	0	2	α	β	0	(R_x, R_y)	(xz, yz)
E_{2g}	2	β	α	0	2	β	α	0		(x^2-y^2, xy)
A_{1u}	1	1	1	1	-1	-1	-1	-1		
A_{2u}	1	1	1	-1	-1	-1	-1	1	z	
E_{1u}	2	α	β	0	-2	$-\alpha$	$-\beta$	0	(x, y)	
E_{2u}	2	β	α	0	-2	$-\beta$	$-\alpha$	0		

D_{6d}	E	$2S_{12}$	$2C_6$	$2S_4$	$2C_3$	$2S_{12}^5$	C_2	$6C_2'$	$6\sigma_d$		
A_1	1	1	1	1	1	1	1	1	1		x^2+y^2, z^2
A_2	1	1	1	1	1	1	1	-1	-1	R_z	
B_1	1	-1	1	-1	1	-1	1	1	-1		
B_2	1	-1	1	-1	1	-1	1	-1	1	z	
E_1	2	$\sqrt{3}$	1	0	-1	$-\sqrt{3}$	-2	0	0	(x, y)	
E_2	2	1	-1	-2	-1	1	2	0	0		(x^2-y^2, xy)
E_3	2	0	-2	0	2	0	-2	0	0		
E_4	2	-1	-1	2	-1	-1	2	0	0		
E_5	2	$-\sqrt{3}$	1	0	-1	$\sqrt{3}$	-2	0	0	(R_x, R_y)	(xz, yz)

D$_{nh}$ Groups

D_{2h}	E	C_2	$C_2(y)$	$C_2(x)$	i	$\sigma(xy)$	$\sigma(xz)$	$\sigma(yz)$		
A_g	1	1	1	1	1	1	1	1		x^2, y^2, z^2
B_{1g}	1	1	-1	-1	1	1	-1	-1	R_z	xy
B_{2g}	1	-1	1	-1	1	-1	1	-1	R_y	xz
B_{3g}	1	-1	-1	1	1	-1	-1	1	R_x	yz
A_u	1	1	1	1	-1	-1	-1	-1		
B_{1u}	1	1	-1	-1	-1	-1	1	1	z	
B_{2u}	1	-1	1	-1	-1	1	-1	1	y	
B_{3u}	1	-1	-1	1	-1	1	1	-1	x	

D_{3h}	E	$2C_3$	$3C_2$	σ_h	$2S_3$	$3\sigma_v$		
A_1'	1	1	1	1	1	1		$x^2 + y^2, z^2$
A_2'	1	1	-1	1	1	-1	R_z	
E'	2	-1	0	2	-1	0	(x, y)	$(x^2 - y^2, xy)$
A_1''	1	1	1	-1	-1	-1		
A_2''	1	1	-1	-1	-1	1	z	
E''	2	-1	0	-2	1	0	(R_x, R_y)	(xz, yz)

D_{4h}	E	$2C_4$	C_2	$2C_2'$	C_2''	i	$2S_4$	σ_h	$2\sigma_v$	$2\sigma_d$		
A_{1g}	1	1	1	1	1	1	1	1	1	1		x^2+y^2, z^2
A_{2g}	1	1	1	-1	-1	1	1	1	-1	-1	R_z	
B_{1g}	1	-1	1	1	-1	1	-1	1	1	-1		x^2-y^2
B_{2g}	1	-1	1	-1	1	1	-1	1	-1	1		xy
E_g	2	0	-2	0	0	2	0	-2	0	0	(R_x, R_y)	(xz, yz)
A_{1u}	1	1	1	1	1	-1	-1	-1	-1	-1		
A_{2u}	1	1	1	-1	-1	-1	-1	-1	1	1	z	
B_{1u}	1	-1	1	1	-1	-1	1	-1	-1	1		
B_{2u}	1	-1	1	-1	1	-1	1	-1	1	-1		
E_u	2	0	-2	0	0	-2	0	2	0	0	(x, y)	

D_{5h}	E	$2C_5$	$2C_5^2$	$5C_2'$	σ_h	$2S_5$	$2S_5^3$	$5\sigma_v$	$\alpha = 2\cos 72°$	$\beta = 2\cos 144°$
A_1'	1	1	1	1	1	1	1	1		x^2+y^2, z^2
A_2'	1	1	1	-1	1	1	1	-1	R_z	
E_1'	2	α	β	0	2	α	β	0	(x, y)	
E_2'	2	β	α	0	2	β	α	0		(x^2-y^2, xy)
A_1''	1	1	1	1	-1	-1	-1	-1		
A_2''	1	1	1	-1	-1	-1	-1	1	z	
E_1''	2	α	β	0	-2	$-\alpha$	$-\beta$	0	(R_x, R_y)	(xz, yz)
E_2''	2	β	α	0	-2	$-\beta$	$-\alpha$	0		

D_{6h}	E	$2C_6$	$2C_3$	C_2	$3C_2'$	$3C_2''$	i	$2S_3$	$2S_6$	$\sigma_h(xy)$	$3\sigma_d$	$3\sigma_v$		
A_{1g}	1	1	1	1	1	1	1	1	1	1	1	1		x^2+y^2, z^2
A_{2g}	1	1	1	1	-1	-1	1	1	1	1	-1	-1	R_z	
B_{1g}	1	-1	1	-1	1	-1	1	-1	1	-1	1	-1		
B_{2g}	1	-1	1	-1	-1	1	1	-1	1	-1	-1	1		
E_{1g}	2	1	-1	-2	0	0	2	1	-1	-2	0	0	(R_x, R_y)	(xz, yz)
E_{2g}	2	-1	-1	2	0	0	2	-1	-1	2	0	0		(x^2-y^2, xy)
A_{1u}	1	1	1	1	1	1	-1	-1	-1	-1	-1	-1		
A_{2u}	1	1	1	1	-1	-1	-1	-1	-1	-1	1	1	z	
B_{1u}	1	-1	1	-1	1	-1	-1	1	-1	1	-1	1		
B_{2u}	1	-1	1	-1	-1	1	-1	1	-1	1	1	-1		
E_{1u}	2	1	-1	-2	0	0	-2	-1	1	2	0	0	(x, y)	
E_{2u}	2	-1	-1	2	0	0	-2	1	1	-2	0	0		

S_n Groups

S_4	E	S_4	C_2	S_4^3		
A	1	1	1	1	R_z	x^2+y^2, z^2
B	1	-1	1	-1	z	x^2-y^2, xy
E	$\begin{cases} 1 \\ 1 \end{cases}$	$\begin{matrix} \imath \\ -\imath \end{matrix}$	$\begin{matrix} -1 \\ -1 \end{matrix}$	$\begin{matrix} -\imath \\ \imath \end{matrix}$	$x+\imath y, R_x+\imath R_y$ $x-\imath y, R_x\imath-\imath R_y$	(xz, yz)

S_6	E	C_3	C_3^2	i	S_6^5	S_6	$\epsilon = \exp(2\pi i/3)$	
A_g	1	1	1	1	1	1	R_z	$x^2+y^2,\ z^2$
E_g	$\begin{matrix}1\\1\end{matrix}$	$\begin{matrix}\epsilon\\\epsilon^*\end{matrix}$	$\begin{matrix}\epsilon^*\\\epsilon\end{matrix}$	$\begin{matrix}1\\1\end{matrix}$	$\begin{matrix}\epsilon\\\epsilon^*\end{matrix}$	$\begin{matrix}\epsilon^*\\\epsilon\end{matrix}$	R_x+iR_y , R_x-iR_y	$(x^2-y^2,\ xy)(xz,\ yz)$
A_u	1	1	1	-1	-1	-1	z	
E_u	$\begin{matrix}1\\1\end{matrix}$	$\begin{matrix}\epsilon\\\epsilon^*\end{matrix}$	$\begin{matrix}\epsilon^*\\\epsilon\end{matrix}$	$\begin{matrix}-1\\-1\end{matrix}$	$\begin{matrix}-\epsilon\\-\epsilon^*\end{matrix}$	$\begin{matrix}-\epsilon^*\\-\epsilon\end{matrix}$	$x+iy$, $x-iy$	

S_8	E	S_8	C_4	S_8^3	C_2	S_8^5	C_4^3	S_8^7	$\epsilon = \exp(2\pi i/8)$	
A	1	1	1	1	1	1	1	1	R_z	$x^2+y^2,\ z^2$
B	1	-1	1	-1	1	-1	1	-1	z	
E_1	$\begin{matrix}1\\1\end{matrix}$	$\begin{matrix}\epsilon\\\epsilon^*\end{matrix}$	$\begin{matrix}i\\-i\end{matrix}$	$\begin{matrix}-\epsilon^*\\-\epsilon\end{matrix}$	$\begin{matrix}-1\\-1\end{matrix}$	$\begin{matrix}-\epsilon\\-\epsilon^*\end{matrix}$	$\begin{matrix}-i\\i\end{matrix}$	$\begin{matrix}\epsilon^*\\\epsilon\end{matrix}$	$x+iy$, $x-iy$	
E_2	$\begin{matrix}1\\1\end{matrix}$	$\begin{matrix}i\\-i\end{matrix}$	$\begin{matrix}-1\\-1\end{matrix}$	$\begin{matrix}-i\\i\end{matrix}$	$\begin{matrix}1\\1\end{matrix}$	$\begin{matrix}i\\-i\end{matrix}$	$\begin{matrix}-1\\-1\end{matrix}$	$\begin{matrix}-i\\i\end{matrix}$		$(x^2-y^2,\ xy)$
E_3	$\begin{matrix}1\\1\end{matrix}$	$\begin{matrix}-\epsilon\\-\epsilon^*\end{matrix}$	$\begin{matrix}i\\-i\end{matrix}$	$\begin{matrix}\epsilon^*\\\epsilon\end{matrix}$	$\begin{matrix}-1\\-1\end{matrix}$	$\begin{matrix}\epsilon\\\epsilon^*\end{matrix}$	$\begin{matrix}-i\\i\end{matrix}$	$\begin{matrix}-\epsilon^*\\-\epsilon\end{matrix}$	R_x+iR_y , R_x-iR_y	$(xz,\ yz)$

Cubic Groups

T

T	E	$4C_3$	$4C_3^2$	$3C_2$		$\epsilon = \exp(2\pi i/3)$
A	1	1	1	1		$x^2+y^2+z^2$
E	1	ϵ	ϵ^*	1		$(2z^2-x^2-y^2,\ x^2-y^2)$
	1	ϵ^*	ϵ	1		
T	3	0	0	-1	$(x,y,z),(R_x,R_y,R_z)$	(xy,xz,yz)

T_h

T_h	E	$4C_3$	$4C_3^2$	$3C_2$	i	$4S_6^5$	$4S_6$	$3\sigma_h$		$\epsilon = \exp(2\pi i/3)$
A_g	1	1	1	1	1	1	1	1		$x^2+y^2+z^2$
E_g	1	ϵ	ϵ^*	1	1	ϵ^*	ϵ	1		$(2z^2-x^2-y^2,\ x^2-y^2)$
	1	ϵ^*	ϵ	1	1	ϵ	ϵ^*	1		
T_g	3	0	0	-1	3	0	0	-1	(R_x,R_y,R_z)	(xy,xz,yz)
A_u	1	1	1	1	-1	-1	-1	-1		
E_u	1	ϵ	ϵ^*	1	-1	$-\epsilon^*$	$-\epsilon$	-1		
	1	ϵ^*	ϵ	1	-1	$-\epsilon$	$-\epsilon^*$	-1		
T_u	3	0	0	-1	-3	0	0	1	(x,y,z)	

T_d / O

T_d	E	$8C_3$	$6\sigma_d$	$6S_4$	$3C_2$		
O	E	$8C_3$	$6C_2'$	$6C_4$	$3C_2(\equiv C_4^2)$		
A_1	1	1	1	1	1		$x^2+y^2+z^2$
A_2	1	1	-1	-1	1		
E	2	-1	0	0	2		$(2z^2-x^2-y^2,\ x^2-y^2)$
T_1	3	0	-1	1	-1	$(x,y,z)(R_x,R_y,R_z)$	(R_x,R_y,R_z)
T_2	3	0	1	-1	-1	(x,y,z) in T_d	(xy,xz,yz)

O_h	E	$8C_3$	$6C_2$	$6C_4$	$3C_2(\equiv C_4^2)$	i	$6S_4$	$8S_6$	$3\sigma_h$	$6\sigma_d$		
A_{1g}	1	1	1	1	1	1	1	1	1	1		$x^2+y^2+z^2$
A_{2g}	1	1	-1	-1	1	1	-1	1	1	-1		
E_g	2	-1	0	0	2	2	0	-1	2	0		$(2z^2-x^2-y^2, x^2-y^2)$
T_{1g}	3	0	-1	1	-1	3	1	0	-1	-1	(R_x, R_y, R_z)	
T_{2g}	3	0	1	-1	-1	3	-1	0	-1	1		(xz, yz, xy)
A_{1u}	1	1	1	1	1	-1	-1	-1	-1	-1		
A_{2u}	1	1	-1	-1	1	-1	1	-1	-1	1		
E_u	2	-1	0	0	2	-2	0	1	-2	0		
T_{1u}	3	0	-1	1	-1	-3	-1	0	1	1	(x, y, z)	
T_{2u}	3	0	1	-1	-1	-3	1	0	1	-1		

Linear Groups

$C_{\infty v}$	E	$2C_\infty(\phi)$	$2C_\infty(2\phi)$	$\cdots$	$\infty\sigma_v$		
$A_1(\Sigma^+)$	1	1	1	$\cdots$	1	z	$x^2+y^2,\ z^2$
$A_2(\Sigma^-)$	1	1	1	$\cdots$	-1	R_z	
$E_1(\Pi)$	2	$2\cos\phi$	$2\cos 2\phi$	$\cdots$	0	$(x,y),(R_x,R_{\cdot y})$	(xz,yz)
$E_2(\Delta)$	2	$2\cos 2\phi$	$2\cos 4\phi$	$\cdots$	0		(x^2-y^2,xy)
$\vdots$	$\vdots$	$\vdots$	$\vdots$		$\vdots$		
					$\vdots$		
$E_n(\Delta)$	2	$2\cos n\phi$	$2\cos 2n\phi$	$\cdots$	0		

$D_{\infty h}$	E	$2C_\infty(\phi)$	$\cdots$	$\infty\sigma_v$	i	$2S_\infty(\phi)$	$\cdots$	$\infty C_2'$		
$A_{1g}(\Sigma_g^+)$	1	1	$\cdots$	1	1	1	$\cdots$	1		$x^2+y^2,\ z^2$
$A_{2g}(\Sigma_g^-)$	1	1	$\cdots$	-1	1	1	$\cdots$	-1	R_z	
$E_{1g}(\Pi_g)$	2	$2\cos\phi$	$\cdots$	0	2	$-2\cos\phi$	$\cdots$	0	(R_x,R_y)	(xz,yz)
$E_{2g}(\Delta_g)$	2	$2\cos 2\phi$	$\cdots$	0	2	$2\cos 2\phi$	$\cdots$	0		(x^2-y^2,xy)
$\vdots$						$\vdots$		$\vdots$		
E_{ng}	2	$2\cos n\phi$	$\cdots$	0	2	$(-1)^n 2\cos 2\phi$	$\cdots$	0		
$\vdots$						$\vdots$		$\vdots$		
$A_{1u}(\Sigma_u^+)$	1	1	$\cdots$	1	-1	-1	$\cdots$	-1	z	
$A_{2u}(\Sigma_u^-)$	1	1	$\cdots$	-1	-1	-1	$\cdots$	1		
$E_{1u}(\Pi_u)$	2	$2\cos\phi$	$\cdots$	0	-2	$2\cos\phi$	$\cdots$	0	(x,y)	
$E_{2u}(\Delta_u)$	2	$2\cos 2\phi$	$\cdots$	0	-2	$-2\cos 2\phi$	$\cdots$	0		
$\vdots$						$\vdots$		$\vdots$		
E_{nu}	2	$2\cos n\phi$	$\cdots$	0	-2	$(-1)^{n+1}2\cos 2\phi$	$\cdots$	0		
$\vdots$						$\vdots$		$\vdots$		

Icosahedral Group

I_h	E	$12C_5$	$12C_5^2$	$20C_3$	$15C_2$	i	$12S_{10}^3$	$12S_{10}$	$20S_6$	15σ	$\alpha = \frac{1}{2}(1+\sqrt{5})$	$\beta = \frac{1}{2}(1-\sqrt{5})$
A_g	1	1	1	1	1	1	1	1	1	1		$x^2+y^2+z^2$
A_{1g}	3	α	β	0	-1	3	α	β	0	-1	(R_x, R_y, R_z)	
T_{2g}	3	β	α	0	-1	3	β	α	0	-1		
G_g	4	-1	-1	1	0	4	-1	-1	1	0		$\left.\begin{array}{l}2z^2-x^2-y^2 \\ x^2-y^2\end{array}\right.$
H_g	5	0	0	-1	1	5	0	0	-1	1		$\left.\begin{array}{l}xy \\ yz \\ xz\end{array}\right\}$
A_u	1	1	1	1	1	-1	-1	-1	-1	-1		
T_{1u}	3	α	β	0	-1	-3	$-\alpha$	$-\beta$	0	1	(x, y, z)	
T_{2u}	3	β	α	0	-1	-3	$-\beta$	$-\alpha$	0	1		
G_u	4	-1	-1	1	0	-4	1	1	-1	0		
H_u	5	0	0	-1	1	-5	0	0	1	-1		

Appendix 4

Hermitian operators

An operator $\mathbf{T}$ transforms a vector $\vec{u}$ in a vector space into a new vector $\vec{v}$ in that same space. Or,

$$\vec{v} = \mathbf{T}\vec{u} \tag{A4.1}$$

The operator $\mathbf{T}$ is linear if

$$\mathbf{T}(\vec{a} + \vec{b}) = \mathbf{T}\vec{a} + \mathbf{T}\vec{b} \tag{A4.2}$$

$$\mathbf{T}(\alpha\vec{a}) = \alpha\mathbf{T}\vec{a} \tag{A4.3}$$

where $\vec{a}$ and $\vec{b}$ are any vectors in the space and α is a constant.

A linear operator can be expressed in terms of a matrix. This is called the matrix representation of $\mathbf{T}$. The effect of $\mathbf{T}$ on the basis vectors of a space is to form a new basis set. Using Eq. A4.1, we obtain, for the components of $\vec{v}$,

$$v_i = \sum_j A_{ij} u_j \tag{A4.4}$$

where the square matrix A is a matrix representation of the operator $\mathbf{T}$.

For a Hermitian matrix, $\mathbf{H}$, it is true that $\mathbf{H}^\dagger = \mathbf{H}$, where the $\dagger$ symbol means take the complex conjugate of the transpose. Thus, $(\mathrm{H}^\dagger)_{ij} = \mathrm{H}_{ij} = \mathrm{H}^*_{ji}$. A Hermitian operator is one that generates a Hermitian matrix as a transformation matrix.

It is a general truth that for Hermitian operators,

$$\langle \mathbf{Hu}|\mathbf{v}\rangle = \langle \mathbf{u}|\mathbf{Hv}\rangle \tag{A4.5}$$

To show this, write

$$\langle \mathbf{Hu}|\mathbf{v}\rangle = \langle [\mathbf{H}\sum_i u_i e_i] | [\sum_j v_j e_j] \rangle$$

$$= [\sum_{ik} u_i H_{ki} e_k]^* [\sum_j v_j e_j]$$

$$= \sum_{ijk} u_i^* H_{ki}^* v_j e_k^* e_j$$

$$= \sum_{ij} u_i^* v_k H_{ki}^*$$

$$= \sum_{ij} u_i^* H_{ik} v_k \qquad (A4.6)$$

$$= \langle \mathbf{u} | \mathbf{Hv} \rangle \qquad (A4.7)$$

where we have made use of the fact that the basis vectors are orthonormal, $e_i e_j = \delta_{ij}$.

Appendix 5
The Dirac delta function

Physical phenomena often occur for an 'infinitely short' duration or are limited to an 'infinitely small' region of space. We need some sort of function to represent such phenomena. Physicists invented the idea of the 'delta' function (δ function) to solve this problem, but it remained for the mathematicians to sort out the rigor of the function with the 'theory of distributions'. We will only define the δ function and give some of its more useful properties.

The Dirac δ function can be defined by the relations

$$\delta(x) = 0 \qquad if \ x \neq 0 \tag{A5.1}$$

or

$$\int \delta(x)\,\mathrm{d}x = 1 \tag{A5.2}$$

where the region of integration includes $x = 0$. An equivalent definition for an arbitrary function $f(x)$ that is continuous at $x = 0$ is

$$\int f(x)\,\delta(x)\,\mathrm{d}x = f(0) \tag{A5.3}$$

Again, the region of integration includes $x = 0$.

One very frequent usage of the Dirac delta function is when we choose our quantum mechanical state basis functions to be plane waves of the form $(1/\sqrt{2\pi\hbar})\exp^{ipx/\hbar}$, where $p/\hbar$ is the wavevector. The integral of a plane wave over the entire x axis diverges, so that plane waves are not square integrable functions. However, provided that we have the definition

$$\frac{1}{2\pi}\int_{-\infty}^{\infty} e^{iku}\,\mathrm{d}k = \delta(u) \tag{A5.4}$$

then we can recapture the useful property that every potential wavefunction can be expanded in terms of plane wave basis functions.

Some useful properties of the δ function include

$$
\begin{aligned}
\delta(x) &= \delta(-x) \\
\delta'(x) &= -\delta'(-x) \\
x\,\delta(x) &= 0 \\
x\,\delta'(x) &= -\delta(x) \\
\delta(ax) &= \frac{1}{a}\delta(x) \\
\delta(x^2 - a^2) &= \frac{1}{2a}[\delta(x-a) + \delta(x+a)] \\
\int \delta(a-x)\,\delta(x-b)\,\mathrm{d}x &= \delta(a-b) \\
f(x)\,\delta(x-a) &= f(a)\,\delta(x-a)
\end{aligned}
$$

$$(A5.5)$$

where a and b are arbitrary constants and a prime denotes differentiation with respect to the argument.

Appendix 6

Moments of distribution functions

The probability of an event occurring can be characterized by a distribution function. This distribution function may, for example, give us the probability of a certain score occurring on an exam or the probability of finding a grain with a certain diameter in a polycrystal. The distribution function may thus be over a field of discrete variables (exam scores) or continuous variables (grain sizes). Since we will always be interested in distribution functions for the energy in a solid, we will focus on continuous variables.†

Rather than having a functional form for the distribution function, sometimes it is more practical and helpful to have the *moments* of the distribution function. In general, the nth moment of a function $f(x)$ about the origin is

$$\mu^{(n)\prime} = \int_{-\infty}^{\infty} x^n f(x)\, dx \tag{A6.1}$$

Likewise, we can define moments about the mean, $\bar{x}$,

$$\mu^{(n)} = \int_{-\infty}^{\infty} (x - \bar{x})^n f(x)\, dx \tag{A6.2}$$

The moments of a distribution function tell us about the shape and structure of the function. The higher the moment, the more detailed the information about the function (and the more difficult the calculation!). In particular, the various moments are:

$$\mu^{(0)} = \int_{-\infty}^{\infty} f(x)\, dx = 1 \tag{A6.3}$$

Equation A6.3 must be true under all circumstances, since a distribution function must be properly normalized. Likewise,

$$\mu^{(1)} = \int_{-\infty}^{\infty} (x - \bar{x}) f(x)\, dx \tag{A6.4}$$

† Strictly speaking, the energy is a semi-continuous variable, but levels are so closely spaced that we may ignore this distinction.

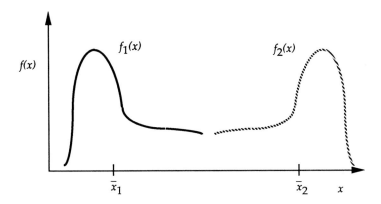

Fig. A6.1. A schematic drawing of a distribution function showing skewness above $(\mu^{(3)} > 0)$ and below $(\mu^{(3)} < 0)$ the average.

since the distribution function is centered about its mean. The second moment,

$$\mu^{(2)} = \int_{-\infty}^{\infty} (x - \bar{x})^2 f(x)\, dx = \mu^{(2)\prime} - \bar{x}^2 \qquad (A6.5)$$

is just the variance of the variable x. The square root of the variance is the standard deviation and, hence, tells us something about the width of the distribution function. The third moment,

$$\mu^{(3)} = \int_{-\infty}^{\infty} (x - \bar{x})^3 f(x)\, dx \qquad (A6.6)$$

tells us about the skewness of the distribution function about the mean. If $\mu^{(3)} > 0$, then $f(x)$ is skewed to values above $\bar{x}$, whereas if $\mu^{(3)} < 0$, $f(x)$ is skewed to values below $\bar{x}$. See Fig. A6.1. Finally,

$$\mu^{(4)} = \int_{-\infty}^{\infty} (x - \bar{x})^4 f(x)\, dx \qquad (A6.7)$$

is a measure of the number of maxima in the function $f(x)$. The coefficient of kurtosis is given by $\mu^{(4)}/(\mu^{(2)})^2$, and a small value means unimodal behavior, while a large value means bimodal behavior.

It is interesting to note that it is sometimes possible to construct a functional form for a distribution function, provided that we know enough of its moments. The reader is referred to advanced statistics and probability textbooks for more knowledge about this process.

Appendix 7

The Sommerfeld expansion

The Sommerfeld expansion is particularly useful when evaluating integrals of the form

$$I \equiv \int_{-\infty}^{\infty} H(\mathscr{E}) f(\mathscr{E}) \, d\mathscr{E} \tag{A7.1}$$

where $f(\mathscr{E})$ is the Fermi–Dirac distribution function,

$$f(\mathscr{E}) = \frac{1}{e^{(\mathscr{E}-\mathscr{E}_F)/k_B T} + 1} \tag{A7.2}$$

and $H(\mathscr{E})$ is any function that vanishes as $\mathscr{E} \to -\infty$ and does not diverge more rapidly than some power of the energy, $\mathscr{E}$.

Now, for a function $H(\mathscr{E})$ which can be written as the first energy derivative of another function,

$$H(\mathscr{E}) = \frac{dG(\mathscr{E})}{d\mathscr{E}} \tag{A7.3}$$

then

$$
\begin{aligned}
I &= \int_{-\infty}^{\infty} \frac{dG(\mathscr{E})}{d\mathscr{E}} f(\mathscr{E}) \, d\mathscr{E} \\
&= G(\mathscr{E}) f(\mathscr{E})|_{-\infty}^{\infty} - \int_{-\infty}^{\infty} G(\mathscr{E}) \frac{\partial f(\mathscr{E})}{\partial \mathscr{E}} \, d\mathscr{E}
\end{aligned} \tag{A7.4}
$$

where the last step follows from integration by parts. The first term on the right-hand side of Eq. A7.4 is zero because $G(\mathscr{E})$ vanishes as $\mathscr{E} \to -\infty$ and the Fermi distribution function vanishes as $\mathscr{E} \to \infty$. The integral I thus has contribution only for energies within a few $k_B T$ of the Fermi energy.

We exploit this property to evaluate the remaining integral in Eq. A7.4. Write $G(\mathscr{E})$ as a Taylor series expansion about the Fermi energy,

$$G(\mathscr{E}) = G(\mathscr{E}_F) + (\mathscr{E} - \mathscr{E}_F) \frac{dG}{d\mathscr{E}}\bigg|_{\mathscr{E}=\mathscr{E}_F} + \frac{1}{2!}(\mathscr{E} - \mathscr{E}_F)^2 \frac{d^2G}{d\mathscr{E}^2}\bigg|_{\mathscr{E}=\mathscr{E}_F} + \dots \tag{A7.5}$$

I can thus be written as a collection of terms of increasing order,

$$I_0 = -G(\mathscr{E}_F) \int_{-\infty}^{\infty} \frac{\mathrm{d}f}{\mathrm{d}\mathscr{E}} \,\mathrm{d}\mathscr{E}$$

$$I_1 = -\left. \frac{\mathrm{d}G}{\mathrm{d}\mathscr{E}} \right|_{\mathscr{E}=\mathscr{E}_F} \int_{-\infty}^{\infty} (\mathscr{E}-\mathscr{E}_F) \frac{\mathrm{d}f}{\mathrm{d}\mathscr{E}} \,\mathrm{d}\mathscr{E}$$

$$I_2 = -\frac{1}{2!} \left. \frac{\mathrm{d}^2 G}{\mathrm{d}\mathscr{E}^2} \right|_{\mathscr{E}=\mathscr{E}_F} \int_{-\infty}^{\infty} (\mathscr{E}-\mathscr{E}_F)^2 \frac{\mathrm{d}f}{\mathrm{d}\mathscr{E}} \,\mathrm{d}\mathscr{E} \qquad (A7.6)$$

and so on for higher-order terms. I_0 is easily evaluated by recognizing that it is the integral over all energies of the distribution function which is equal to 1. I_1 vanishes because it is the integral of an odd function. I_2 can be integrated numerically by letting $x \equiv (\mathscr{E}-\mathscr{E}_F)/k_B T$, so that

$$I_2 = \frac{1}{2!} \left. \frac{\mathrm{d}^2 G}{\mathrm{d}\mathscr{E}^2} \right|_{\mathscr{E}=\mathscr{E}_F} \left((k_B T)^2 \int_{-\infty}^{\infty} \frac{x^2 \mathrm{e}^x}{(1+\mathrm{e}^x)^2} \,\mathrm{d}x \right) \qquad (A7.7)$$

The expression in parentheses is tabulated in mathematical handbooks. Hence, we arrive at

$$I_2 = \frac{\pi^2}{6} (k_B T)^2 \left. \frac{\mathrm{d}^2 G}{\mathrm{d}\mathscr{E}^2} \right|_{\mathscr{E}=\mathscr{E}_F} \qquad (A7.8)$$

Bibliography

The Periodic Table
Puddephatt, R.J. and Monaghan, P.K. (1986). *The Periodic Table of the Elements* (Clarendon Press, Oxford).
Sanderson, R.T. (1960). *Chemical Periodicity* (Chapman & Hall, London).
van Spronsen, J.W. (1969). *The Periodic System of Chemical Elements* (Elsevier, Amsterdam).

General Sources on Bonding
Altmann, S.A. (1991). *Band Theory of Solids: An Introduction from the Point of View of Symmetry* (Clarendon Press, Oxford).
Coulson, C.A. (1952). *Valence* (Clarendon Press, Oxford). A classic book.
Cox, P.A. (1987). *The Electronic Structure and Chemistry of Solids* (Oxford University Press, Oxford). Clear and well-written discussions of most aspects of bonding.
Harrison, W.A. (1980). *Electronic Structure and the Properties of Solids: The Physics of the Chemical Bond* (Freeman, San Francisco).
Hoffmann, R. (1988). *Solids and Surfaces: A Chemist's View of Bonding in Extended Structures* (VCH Publishers, New York). Terse, but comprehensive discussion of bonding from the chemist's viewpoint.
Pauling, L. (1939). *The Nature of the Chemical Bond and the Structure of Molecules and Crystals: An Introduction to Modern Structural Chemistry* (Cornell University Press, Ithaca, NY). Another classic.
Sutton, A.P. (1993). *Electronic Structure of Materials* (Clarendon Press, Oxford). Well-written and comprehensive discussion of bonding from the real-space approach.

Group Theory
Burns, G. (1977). *Introduction to Group Theory with Applications* (Academic Press, New York).
Cotton, F.A. (1990). *Chemical Applications of Group Theory* (Wiley, New York). The easiest and most helpful discussion of group theory from the chemist's viewpoint.
Tinkham, M. (1964). *Group Theory and Quantum Mechanics* (McGraw-Hill, New York).
Wigner, E.P. (1959). *Group Theory and its Application to the Quantum Mechanics of Atomic Spectra* (Academic Press, New York).

303

Quantum Mechanics

Cohen-Tannoudji, C., Diu, B., and Laloë, F. (1977) *Quantum Mechanics* (Wiley, New York). Has everything you ever wanted to know; just difficult to find.

Sakurai, J.J. (1976). *Advanced Quantum Mechanics* (Addison-Wesley, Reading, Mass.). The standard undergraduate textbook.

Schiff, L.I. (1968). *Quantum Mechanics* (McGraw-Hill, New York). The standard graduate textbook.

Condensed Matter Physics

Ashcroft, N.W. and Mermin, N.D. (1976). *Solid State Physics* (Holt, Rinehart and Winston, New York).

Burns, G. (1985). *Solid State Physics* (Academic Press, New York).

Harrison, W.A. (1970). *Solid State Theory* (McGraw-Hill, New York).

Kittel, C. (1986). *Introduction to Solid State Physics* (Wiley, New York).

More Specialized Topics

Altmann, S.L. (1992). *Icons and Symmetries* (Clarendon Press, Oxford). Contains a good chapter on the Peierls instability.

Brillouin, L. (1946). *Wave Propagation in Periodic Structures: Electric Filters and Crystal Lattices* (McGraw-Hill, New York). An engaging little book.

Hammond, C. (1990). *Introduction to Crystallography* (Oxford University Press, Oxford).

Harrison, W.A. (1966). *Pseudopotentials in the Theory of Metals* (W.A. Benjamin, New York).

Moruzzi, V.L., Janak, J.F., and Williams, A.R. (1978). *Calculated Electronic Properties of Metals* (Pergamon Press, New York).

West, A.R. (1988). *Basic Solid State Chemistry* (Wiley, New York).

Zallen, R. (1983). *The Physics of Amorphous Solids* (Wiley, New York).

Index